Heat and Mass Transfer in Buildings

Also available from Taylor & Francis

Energy Management in Buildings
K. Moss Hb: ISBN 9780415353915
Pb: ISBN 9780415353922

Building Services Engineering 5th edition
D. Chadderton Hb: ISBN 9780415413541
Pb: ISBN 9780415413558

Tropical Urban Heat Island
N.H. Wong *et al.* Hb: ISBN 9780415411042

Renewable Energy Resources 2nd edition
J. Twidell and T. Weir Hb: ISBN 9780419253204
Pb: ISBN 9780419253303

Mechanics of Fluids 8th edition
B. Massey, by J. Ward Smith Hb: ISBN 9780415362054
Pb: ISBN 9780415362061

Ventilation Systems
H. Awbi Pb: ISBN 9780419217008

Housing & Asthma
S. Howieson Hb: ISBN 9780415336451
Pb: ISBN 9780415336468

Information and ordering details

For price availability and ordering visit our website **www.tandf.co.uk/builtenvironment**

Alternatively our books are available from all good bookshops.

Heat and Mass Transfer in Buildings

Second edition

Keith J. Moss

Routledge
Taylor & Francis Group
LONDON AND NEW YORK

First published 1998
by Taylor & Francis

Second edition published 2007
by Taylor & Francis
2 Park Square, Milton Park, Abingdon, Oxon, OX14 4RN

Simultaneously published in the USA and Canada
by Taylor & Francis
711 Third Avenue, New York, NY 10017, USA

*Taylor & Francis is an imprint of the Taylor & Francis Group,
an informa business*

© 1998, 2007 Keith J. Moss

Typeset in Sabon by
Newgen Imaging Systems Pvt Ltd, Chennai, India

All rights reserved. No part of this book may be reprinted or
reproduced or utilised in any form or by any electronic,
mechanical, or other means, now known or hereafter
invented, including photocopying and recording, or in any
information storage or retrieval system, without permission in
writing from the publishers.

The publisher makes no representation, express or implied, with
regard to the accuracy of the information contained in this book and
cannot accept any legal responsibility or liability for any efforts or
omissions that may be made.

British Library Cataloguing in Publication Data
A catalogue record for this book is available from the British Library

Library of Congress Cataloging in Publication Data
Moss, Keith.
 Heat and mass transfer in buildings / Keith Moss. – 2nd ed.
  p. cm.
 Includes bibliographical references and index.
  1. Heating–Mathematics. 2. Heat–Transmission–Mathematical
models. 3. Numerical calculations. I. Title.
TH7124.M67 2007
697–dc22   2006034233

ISBN 978–0–415–40907–0 (hbk)
ISBN 978–0–415–40908–7 (pbk)

Contents

List of examples	x
List of case studies	xv
Preface to the second edition	xvii
Acknowledgements	xix
Introduction	xxi

1 Thermal comfort and assessment   1

    1.1 *Introduction 2*
    1.2 *Heat energy and temperature 2*
    1.3 *Thermometry 3*
    1.4 *Types of thermometer 4*
    1.5 *Heat loss from the human body 5*
    1.6 *Physiological responses 10*
    1.7 *Thermal assessment 11*
    1.8 *Thermal comfort criteria 17*
    1.9 *Temperature profiles 24*
    1.10 *Chapter closure 25*

2 Heat conduction   26

    2.1 *Introduction 27*
    2.2 *Heat conduction at right angles to the surface 27*
    2.3 *Surface conductance 31*
    2.4 *Heat conduction in ground floors 36*
    2.5 *Heat conduction in suspended ground floors 38*
    2.6 *Thermal bridging and non-standard U values 41*
    2.7 *Non-standard U values, multi-webbed bridges 43*
    2.8 *Radial conductive heat flow 46*
    2.9 *Chapter closure 53*

3 Heat convection — 54

- 3.1 Introduction *54*
- 3.2 Rational formulae for free and forced heat convection *57*
- 3.3 Temperature definitions *59*
- 3.4 Convective heat output from a panel radiator *61*
- 3.5 Heat output from a freely suspended pipe coil *63*
- 3.6 Heat transfer from a tube in a condensing secondary fluid *64*
- 3.7 Cooling flux from a chilled ceiling *66*
- 3.8 Heat flux off a floor surface from an embedded pipe coil *68*
- 3.9 Heat transfer notes *70*
- 3.10 Chapter closure *71*

4 Heat radiation — 72

- 4.1 Introduction *73*
- 4.2 Surface characteristics *73*
- 4.3 The greenhouse effect *76*
- 4.4 Spectral wave forms *76*
- 4.5 Monochromatic heat radiation *77*
- 4.6 Laws of black body radiation *78*
- 4.7 Laws of grey body radiation *80*
- 4.8 Radiation exchange between a grey body and a grey enclosure *81*
- 4.9 Heat transfer coefficients for black and grey body radiation *82*
- 4.10 Heat radiation flux I *83*
- 4.11 Problem solving *84*
- 4.12 Asymmetric heat radiation *96*
- 4.13 Historical references *97*
- 4.14 Chapter closure *97*

5 Measurement of fluid flow — 98

- 5.1 Introduction *98*
- 5.2 Flow characteristics *99*
- 5.3 Conservation of energy in a moving fluid *100*
- 5.4 Measurement of gauge pressure with an uncalibrated manometer *101*

5.5 *Measurement of pressure difference with an uncalibrated differential manometer 102*
5.6 *Measurement of flow rate using a venturi meter and orifice plate 104*
5.7 *Measurement of air flow using a pitot static tube 111*
5.8 *Chapter closure 114*

6 Characteristics of laminar and turbulent flow   115

6.1 *Introduction 115*
6.2 *Laminar flow 116*
6.3 *Turbulent flow 119*
6.4 *Boundary layer theory 121*
6.5 *Characteristics of the straight pipe or duct 125*
6.6 *Determination of the frictional coefficient in turbulent flow 126*
6.7 *Solving problems 127*
6.8 *Chapter closure 135*

7 Mass transfer of fluids in pipes, ducts and channels   136

7.1 *Introduction 137*
7.2 *Solutions to problems in frictionless flow 137*
7.3 *Frictional flow in flooded pipes and ducts 144*
7.4 *Semi-graphical solutions to frictional flow in pipes and ducts 160*
7.5 *Gravitational flow in flooded pipes 162*
7.6 *Gravitational flow in partially flooded pipes and channels 170*
7.7 *Alternative rational formulae for partial flow 176*
7.8 *Flow of natural gas in pipes 180*
7.9 *Flow of compressed air in pipes 181*
7.10 *Vacuum pipe sizing 183*
7.11 *Chapter closure 184*

8 Natural ventilation in buildings   185

8.1 *Introduction 186*
8.2 *Aerodynamics around a building 186*
8.3 *Effects on cross-ventilation from the wind 191*
8.4 *The stack effect 194*

viii   Contents

    8.5 *Natural ventilation to internal spaces with openings in one wall only 198*
    8.6 *Ventilation for cooling purposes 200*
    8.7 *Fan assisted ventilation 205*
    8.8 *Further reading 206*
    8.9 *Chapter closure 206*

9 Regimes of fluid flow in heat exchangers   207

    9.1 *Introduction 208*
    9.2 *Parallel flow and counterflow heat exchangers 209*
    9.3 *Heat transfer equations 212*
    9.4 *Heat exchanger performance 219*
    9.5 *Cross flow 225*
    9.6 *Further examples 228*
    9.7 *Chapter closure 232*

Appendix 1: verifying the form of an equation by dimensional analysis   233

    A1.1 *Introduction 233*
    A1.2 *Dimensions in use 234*
    A1.3 *Appendix closure 238*

Appendix 2: solving problems by dimensional analysis   239

    A2.1 *Introduction 240*
    A2.2 *Establishing the form of an equation 240*
    A2.3 *Dimensional analysis in experimental work 243*
    A2.4 *Examples in dimensional analysis 244*
    A2.5 *Appendix closure 262*

Appendix 3: renewable energy systems   263

    A3.1 *Introduction 263*
    A3.2 *Wind turbines 264*
    A3.3 *Hydro power 267*
    A3.4 *Marine turbines 275*
    A3.5 *Solar irradiation and the solar constant 277*
    A3.6 *Photovoltaics 281*
    A3.7 *Biomass 282*
    A3.8 *Combined heat and power 285*
    A3.9 *Fuel cell CHP 287*

A3.10   References and further reading 289
A3.11   Appendix closure 290

Appendix 4: towards sustainable building engineering 291

A4.1   Introduction 291
A4.2   Thermodynamics and sustainability 292
A4.3   The laws of thermodynamics 294
A4.4   Power supplies 297
A4.5   Products and systems 297
A4.6   The building footprint 300
A4.7   Scenarios for building services 300
A4.8   Further reading 302
A4.9   Appendix closure 303

Bibliography   304
Index   305

Examples

1.1	Determination of equations for comfort and mean radiant temperature	14
1.2	Determination of required ambient temperature for an office	18
1.3	Determination of mean radiant temperature for an occupied store room	19
1.4	Evaluate the conditions that provide thermal comfort for spectators	19
1.5	Finding the ambient temperature required for store assistants	19
1.6	Changing air temperature controls to maintain thermal comfort	21
1.7	Recommending ambient conditions for factory operatives	21
1.8	Determination of the predicted percentage dissatisfied (PPD)	24
2.1	Conductive heat flow through an external wall	32
2.2	Finding the conductive heat flux and interface temperatures through an external wall	33
2.3	Calculating three thermal transmittance values through a ground floor	37
2.4	Determination of the thermal transmittance for a suspended ground floor	39
2.5	Calculating the non-standard transmittance through a bridged external wall	42
2.6	Determination of the non-standard transmittance through a hollow concrete block	44
2.7	Calculating the rate of heat loss from a cylindrical thermal storage vessel	47
2.8	Determining the rate of heat loss from an insulated steam pipe	49
2.9	Calculating the thickness of thermal insulation for a pipe	50
2.10	Determining the minimum thickness of thermal insulation required to prevent surface condensation	52
3.1	Calculating mean bulk and mean film temperature, and mean temperature difference for a given rise in temperature of the secondary fluid	59

3.2	Calculating mean bulk temperature and mean film temperature for primary and secondary fluids	60
3.3	Finding the LMTD in counterflow	60
3.4	Determining the convective heat transfer from a panel radiator	61
3.5	Calculating the heat output from a greenhouse pipe coil	63
3.6	Determination of the convective heat transfer coefficient and surface area for a water cooled condenser	64
3.7	Determination of the cooling flux from a chilled beam ceiling	66
3.8	Calculation of the heat flux from the surface of a floor containing embedded pipe coils	68
3.9	Determining the length of a single pass shell and tube condenser	69
4.1	Calculation of the radiant emission from a luminous quartz heater	84
4.2	Determination of the effect of a bright aluminium foil located in an external cavity wall	85
4.3a	Determination of the effect of a bright metal foil located behind a radiator	88
4.3b	Calculation of the upward and downward emission from a radiant panel	90
4.4	Calculation of the effects on heat emission from a steam pipe having different thermal insulation applications	92
4.5	Determination of the efficiency of a gas fired radiant heater	94
4.6	Determination of the convection coefficient between flue gas and stem thermometer	94
5.1	Determining static pressure of air using a manometer	101
5.2	Determining whether an air filter needs replacing using a differential manometer	103
5.3	Find the constant and the flow rate of water using an inclined differential manometer	107
5.4	Calculate the rate of steam flow using an orifice plate	108
5.5	Sizing a recording chart for a venturi meter fitted to a water main	108
5.6	Determine the volume flow rate of air in a duct using a pitot static tube	113
6.1	Calculate the pressure loss of water flowing in a pipe	128
6.2	Determine the pumping power required and maximum laminar flow rate of oil in a pipeline	129
6.3	Show that maximum velocity is twice the mean in laminar flow through a pipe	131
6.4	Calculate the diameter of the rising main that serves two high level tanks	132
6.5	Determine the static pressure loss of air flowing in a duct	134

7.1	Find the maximum length of a pump suction pipe set to a gradient	137
7.2	Determine the mass transfer of water passing through a syphon pipe	139
7.3	Calculate the gauge pressure at a hydrant valve when it is shut and when it is open	141
7.4	Find the diameter at the larger end of a transformation piece through which air is flowing	142
7.5	Determine the power available at the nozzle of an open fire hydrant valve	144
7.6	Calculate the shock loss as water flows through a sudden enlargement	147
7.7	Determine the shock loss and mass transfer as oil passes through a sudden enlargement	148
7.8	Calculate the shock loss and mass transfer as water passes through a sudden contraction	150
7.9	Determine the velocity pressure loss factor for a globe valve	152
7.10	Calculate the velocity head loss factor for a gate valve	153
7.11	Calculate the pressure loss through part of a space heating system	153
7.12	Determine the head loss and gradient as water passes through a pipeline	154
7.13	Calculate the static regain in a transformation piece carrying air	155
7.14	Determine the duty and output power of a booster pump delivering mains water from a ground storage tank	157
7.15	Sizing the rising main to two high level water storage tanks	159
7.16	Calculate the pressure loss in a horizontal pipe	160
7.17	Determine the specific static pressure loss in a circular duct	162
7.18	Calculate the gravitational mass transfer of water in a connecting pipe	163
7.19	Calculate the total gravitational mass transfer of water in two pipes	164
7.20	Determine the gravitational mass transfer of water in a vertical pipe	165
7.21	Determine the gravitational mass transfer of water in two pipes from a common pipeline	167
7.22	Calculate the discharge capacity of a drain pipe set to a gradient	171
7.23	Determine the required gradient and mass transfer of soil water in a drain pipe	171
7.24	Calculate the size of a vertical stack, given simultaneous flow	172
7.25	Size the rain water drain pipes serving a car park	173
7.26	Determine the mass transfer of water in an open channel	175

7.27	Compare the mean fluid velocity in a 100 mm bore pipe set to a gradient using four rational formulae	177
7.28	Compare the mean water velocity and mass transfer in a channel set to a gradient using four rational formulae	177
7.29	Verifying Pole's formula for sizing a pipe carrying natural gas	181
7.30	Sizing a compressed air pipeline	182
7.31	Sizing a pipeline under vacuum	184
8.1	Calculate the wind pressure on a building facade	191
8.2	Determine the minimum ventilation rate due to wind speed	192
8.3	Calculate the stack effect in a multi-storey building	196
8.4	Determine the ventilation rate due to stack effect	196
8.5	Calculate the volume flow of air into an internal space due to an opening on the windward side	199
8.6	Determine the volume flow and air change into a room from a partially open casement window	199
9.1	Find the log mean temperature difference	211
9.2	Compare the advantages of counterflow over parallel flow	215
9.3	Determine the heating surface for a shell and tube heat exchanger	216
9.4	Calculate the heating surface and output for a vapour compression refrigeration condenser	218
9.5	Determine the effectiveness and fluid outlet temperatures of an economiser	221
9.6	Calculate the heat exchange surface, capacity ratio, effectiveness and number of transfer units for an HWS calorifier	224
9.7	Determine the heat exchange surface, capacity ratio, effectiveness and number of transfer units for a cross flow air heater battery	226
9.8	Calculate the mass transfer of the secondary fluid and the length of the primary tube bundle for a non-storage heating calorifier	228
9.9	Determine the condenser output, the leaving temperature of the coolant and the length of the tube bundle for a vapour compression refrigeration unit	231
A1.1	Verify the equation for pressure	235
A1.2	Verify the D'Arcy equation	236
A1.3	Show that the Reynolds number is dimensionless	236
A1.4	Verify Box's formula	236
A1.5	Verify the mass transfer equation	237
A1.6	Verify the Bernoulli equation	237
A1.7	Verify the dimensions for the Stefan–Boltzman constant for heat radiation	237

A2.1	Verify the centrifugal pump and fan laws	244
A2.2	Show the dimensionless groups in an equation for the power of a fan	247
A2.3	Find the scale and corresponding speed of a model for a given fan performance	248
A2.4	Verify the dimensionless groups for the wall shear stress for the mass transfer of fluid in a pipe	250
A2.5	Find the form of the equation for forced convection from a fluid transported in a straight pipe	251
A2.6	Find the form of the equation for free convection in turbulent flow over vertical plates	254
A2.7	Find the form of an equation for the rise in surface temperature of a wall, the dimensionless groups in the equation and the heat up time	256
A2.8	Verify the form of Pole's formula for natural gas flow in a pipe	261
A3.1	Find the annual energy generated from a wind farm, the surplus energy available to the National Grid, and the carbon dioxide emissions saved	265
A3.2	Determine the flow of water over a weir to generate 25 kW	267
A3.3	Find the mass transfer of water into a hydro turbine, the turbine power generated and the energy stored in the upper reservoir	270
A3.4	Find the gravitational mass transfer of water handled by a hydro turbine and the power developed and calculate the pump duty to return the water to the upper storage reservoir	272
A3.5	Find the potential power output and annual energy output from an array of marine turbines	276
A3.6	Find the rate of energy collection from solar thermal collectors and the collection efficiency	280
A3.7	Find the rate of energy absorption from solar thermal collectors	280

Case studies

1.1	Evaluation of comfort and mean radiant temperatures for sedentary occupation	15
8.1	Size the free area for openings, estimate the rate of cooling and daily cooling energy extraction for a five-storey building	200
A2.1	Verify the form of the D'Arcy equation	240
A2.2	Using a model to verify the pressure loss in a full size counterpart	243

Preface to the second edition

This is the last second edition in the trilogy. In this book, I have endeavoured to provide text, problems and solutions that relate the subjects of Heat and Mass Transfer to the discipline of Building Engineering.

Essentially, buildings are subject to air movement due to the outdoor and the indoor climate. Use is made of the mass transfer of water, refrigerants, air and sometimes steam to maintain the indoor climate. The presence, or otherwise, of the building occupants influences the indoor climate as well. Finally, the building envelope acts as a climate diurnal time capsule for the benefit of its occupants.

In this new edition, two new appendices have been included. Appendix 3 – Renewable Energy Systems – includes some of the fuels and technologies currently available to building engineers. Appendix 4 – Towards Sustainable Building Engineering – introduces the role that thermodynamics has in understanding why building engineering must go sustainable, and offers two scenarios to emphasise the importance of the changes that building engineers face in the design, erection and installation, and operating life of buildings.

Solar irradiation and solar collectors have been moved from Chapter 4 to Appendix 3. Chapter 7 has been extended to include the application of rational formulae applied to pipe sizing natural gas, compressed air and vacuum. Chapters 10 and 11 on dimensional analysis have been moved to Appendices 1 and 2 as the subject, although important, is more academic than practical.

Some of the questions in the first edition have been revisited to make them more meaningful.

A total of 11 new worked examples have been included in this edition; three in Chapter 7, and eight in the Appendices.

While there is currently a shift towards making people multi-skilled, this does not infer that programmes of learning can become generic. Indeed, it is increasingly the case that courses are now more specific to the needs of the individual at the workplace. At least one national awarding authority has emphasised for some time that their programmes of study must be related to workplace activities.

This provides a particular challenge to lecturers, teachers and authors who deliver ancillary subjects, such as mathematics, engineering science and thermofluids. I strongly feel that as a practitioner, lecturer and author, I have a commitment to ensure that the learning experience is rich and rewarding, and this in my view is primarily achieved through making ancillary subjects which underpin the primary subjects of a course of study, current and relevant.

Acknowledgements

I acknowledge with thanks the permission granted by the Chartered Institution of Building Services Engineers to reproduce some data from the *CIBSE Guides*. Also, my thanks is extended to the Heating and Ventilating Contractors Association for permission to use material in Chapter 8 from the open learning publications. The material used in this book has been prepared and amended over many years. If copyright permission has been overlooked, the publishers will be pleased to take up the matter on request.

I have to acknowledge the patience of my teachers. In particular, my thanks are due to Alec Griffiths who taught me the basic rules of mathematics and then showed me how mathematics can be used as a tool for solving practical problems. Another person whose name escapes me was an Australian lecturer at what was then the Southbank Polytechnic, who in 1960, with his warmth of personality and extraordinary ability to teach, inspired the whole group of us mathematically ignorant students, so that by the end of the course we were sufficiently competent, and motivated, to successfully pass the examination. There are two other people that deserve acknowledgement here; they are Tony Barton with whom I worked in the teaching business for 25 years and Anne Noblett from HVCA with whom I worked on open learning material for the HND.

Introduction

This book is intended to provide, within the limits of its title, the underpinning knowledge for students studying subjects in building engineering.

The reader will find that it is necessary to participate and respond to the narrative which has been written for those with a good grounding in mathematics who have an interest, vested or otherwise, in building engineering technology.

With the explosion of IT in the form of dedicated software for design purposes it is very easy not to give sufficient attention to fundamental theory and even design calculations. A balance in the process of course-delivery has somehow to be struck between the acquisition of underpinning knowledge and developing the skills required to use dedicated software and computer aided design systems.

One way to achieve a balance for the student is to ensure that the support subjects, such as heat and mass transfer, are dedicated to relating fundamental principles to practical design applications. This will help to secure an interest at least in an important part of the learning process.

If after reading and participating in parts or all of this book, it has provided a learning experience and the reader has been enthused by even a little, my efforts will have been rewarded.

Chapter 1

Thermal comfort and assessment

Nomenclature

A	area (m^2)
ASHRAE	American Society of Heating Refrigeration and Air Conditioning Engineers
Clo	unit of thermal resistance of clothing (m^{2}K/W)
dt	temperature difference (K)
eh_r	heat transfer coefficient for radiation (W/m^{2}K)
emf	electromotive force (volts)
h	height (m)
h_c	heat transfer coefficient for convection (W/m^{2}K)
m	mass (kg)
Met.	metabolic rate (W/m^2) of body surface
mwet	mean weighted enclosure temperature (°C)
P	pressure (Pa)
PMV	predicted mean vote
PPD	predicted percentage dissatisfied
Q	rate of heat loss (W)
Q_c	rate of heat loss/gain by convection (W)
Q_{cd}	rate of heat loss/gain by conduction (W)
Q_e	rate of heat loss by evaporation (W)
Q_r	rate of heat loss/gain by radiation (W)
t	temperature (°C)
t_a	air temperature (°C)
t_{am}	ambient temperature (°C)
t_c	comfort temperature (°C)
t_e	environmental temperature (°C)
t_g	globe temperature (°C)
t_r	mean radiant temperature (°C)
u	mean air velocity (m/s)
V	volume (m^3)

1.1 Introduction

This first chapter introduces you to temperature, the variations of which provides the motive force in heat transfer, and heat energy, the flow and transport of which in air, water and steam is the essence of much of heating, ventilating and air conditioning design. Its main focus, however, is on the topic of thermal comfort and the assessment of indoor climates, in which people live and work, to establish levels of comfort. The American Society of Heating Refrigeration and Air Conditioning Engineers (ASHRAE) defines thermal comfort as 'that condition of mind in which satisfaction is expressed with the thermal environment'. The accurate assessment of building heat losses and gains, the type of comfort systems selected and the regimes of control of the comfort systems are all directed towards achieving this definition.

The effect that the amount of clothing, which is worn, has on different levels of activity also impinges on the comfort of the individual. Because thermal comfort is also a subjective assessment, a minority of individuals may feel uncomfortable even in thermal environments which are well regulated.

1.2 Heat energy and temperature

A definition of energy is the capacity a substance possesses which can result in the performance of work. It is a property of the substance. Heat on the other hand is energy in transition. Heat is one form of energy and can be expressed, for example, as a specific heat capacity in kJ/kgK. In this form, it is expressing the potential of a substance for storing heat which it has absorbed from its surroundings. It can also express the potential for the intensity of heat transfer from the substance to its surroundings.

Up until the end of the eighteenth century, heat energy, known as 'caloric', was considered as a fluid which could be made to flow for the purposes of space heating among other things or it flowed of its own volition as a result of friction which was generated as a result of a process or work done such as boring out a cannon. The idea of heat being a form of energy rather than a fluid was developed by an American named Benjamin Thompson, subsequently known as Count Rumford, during the process of boring out cannons for his arsenal as war minister of Bavaria. His conclusions were that the amount of heat liberated depended upon the work done against friction by the boring device. A partial definition of heat energy is, therefore, the interaction between two substances which occurs by virtue of their temperature difference when they communicate. However, heat energy does not always initiate a rise in temperature as in the cases of the latent heat of vaporisation and condensation which occur when substances change in state. This is a qualification of the definition.

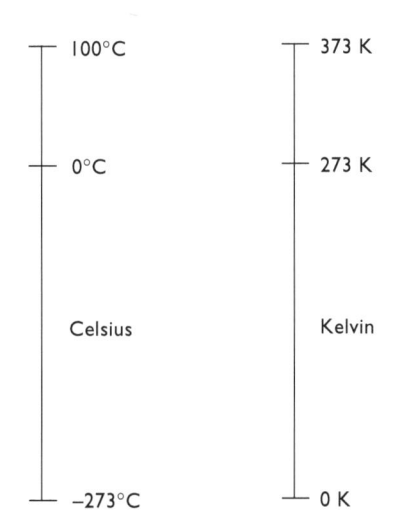

Figure 1.1 Scales of temperatures.

Heat is a transient commodity like work; it exists during communication only, although like work its effect may be permanent. The primary need for burning fuel oil might be the generation of heat energy and the permanent result of the process are the products of combustion. The combustion products cannot return to fuel oil. The transient result is the generation of heat.

A definition of temperature is a scaled measurement of relative hot and cold sensations. It can be described as an intensity of hotness or coldness. Kelvin found that absolute coldness is reached when the agitation of the molecules and atoms of a substance ceases at $-273.15°C$ (0.0 K). Temperature scales have been advanced by various authorities. The scales commonly in use now are the Celsius scale and the Kelvin scale. Figure 1.1 shows these scales from absolute zero to the upper fixed point.

1.3 Thermometry

In the seventeenth century it was proposed that two fixed points should be used to determine a temperature scale:

- The lower fixed point was taken as melting ice at atmospheric pressure; the ice being distilled water. The ice point is the temperature at which ice and water can exist in equilibrium.
- The upper fixed point was taken as steam generated from distilled water when boiling at atmospheric pressure. Since the temperature tends to

4 Thermal comfort and assessment

Table 1.1 Fixed points of the International Temperature Scale

Fixed points	Temperature (°C)
Boiling point of liquid oxygen	−182
Ice point	0
Steam point	100
Boiling point of sulphur	444.6
Freezing point of silver	960.8
Freezing point of gold	1063.0

vary depending upon geographical location, the steam point is the temperature of boiling water and steam at atmospheric pressure on latitude 45.

Table 1.1 lists the fixed points of the International Temperature Scale at standard atmospheric pressure (101 325 Pa).

1.4 Types of thermometer

There are six main types of thermometer, namely

1 constant volume gas thermometer,
2 resistance thermometers,
3 thermocouples,
4 liquid thermometers,
5 bimetallic thermometer,
6 pyrometers.

The constant volume gas thermometer was selected in 1887 as the standard: it did not give a pointer reading, however. For a perfect gas, Boyle's Law states that $P \propto 1/V$ at constant temperature and the Kelvin scale (Figure 1.1) agrees exactly with the scale of a perfect gas thermometer.

The resistance thermometer consists of a platinum wire wound onto two strips of mica and the coil is attached to leads, which is connected in turn to a wheatstone bridge.

Thermocouples have the measuring element as the junction between two dissimilar metal wires. The emf generated at the junction results from its temperature and this is measured accurately on a potentiometer or approximately using a galvanometer.

The liquid thermometer relies on the expansion of a liquid in a glass or steel tube in response to a rise in temperature.

The bimetallic thermometer is associated with dial instruments in which two dissimilar metal strips soldered together in a coil expand differentially on rise in temperature, thus, moving a pointer round the dial.

Pyrometers, of which there are four types, tend to be used for measuring very low temperatures and temperatures up to 1100°C. The four types are: resistance, thermoelectric, radiation and optical.

1.5 Heat loss from the human body

The core temperature of the human body is taken as 37.2°C. This implies that humans, as with all mammals, must generate heat in order to maintain body temperature. About 80% of food intake is required to maintain the body temperature. The metabolic rate refers to the rate at which energy is released from food into the body cells. It is affected by a person's size, body fat, sex, age, hormones and level of activity. See Table 1.2 which lists the approximate total heat output for different levels of activity. The Met. is equal to the metabolic rate for a seated adult at rest and is equivalent to 58 W/m^2 of body surface.

Table 1.3 lists the metabolic rate in units of Met. and W/m^2 for different levels of activity.

In order to estimate the heat loss from a person's body the surface area of the body is required. A close approximation of surface area A is obtained from the person's height h and mass m using an empirical formula attributed to Dubois where:

$$A = (m^{0.425} \times h^{0.725} \times 0.2024) \text{ m}^2$$

Thus, for a person of mass 70 kg and height 1.8 m,

$$A = 6.1 \times 1.53 \times 0.2024 = 1.89 \text{ m}^2$$

A figure of 2 m^2 is frequently used for the surface area of a clothed adult. Discomfort is felt if body temperature varies much from the core temperature of 37.2°C. A loss of about 2.5 K for an extended period may induce a state of hypothermia. Particularly, in the old and the young this is a serious condition leading ultimately to death.

A state of hypothermia means that the body cannot restore itself to its normal temperature without the aid of active heating, for example, being put into a pre-warmed bed which is then maintained at a constant temperature. The intake of high calorie hot food will also be necessary to aid the recovery process. Wrapping the sufferer in warm and reflective clothing may be a temporary measure but not one which will restore the patient's body temperature.

If body temperature rises by 4 K for an extended period, a state of hyperpyrexia may be induced. In this condition, the body is unable to liberate

Table 1.2 Heat emission from the human body (adult male, body surface area 2 m^2)

Application			Sensible (s) and latent (l) heat emissions/W at the stated dry-bulb temperatures/°C									
Degree of activity	Typical	Total	15		20		22		24		26	
			(s)	(l)	(s)	(l)	(s)	(l)	(s)	(l)	(s)	(l)
Seated at rest	Theatre, hotel lounge	115	100	15	90	25	80	35	75	40	65	50
Light work	Office, restaurant*	140	110	30	100	40	90	50	80	60	70	70
Walking slowly	Store, bank	160	120	40	110	50	100	60	85	75	75	85
Light bench work	Factory	235	150	85	130	105	115	120	100	135	80	155
Medium work	Factory, dance hall	265	160	105	140	125	125	140	105	160	90	175
Heavy work	Factory	440	200	220	190	250	165	275	135	305	105	335

Source: Reproduced from the *CIBSE Guide* (1986) by permission of the Chartered Institute of Building Services Engineers.

Note

* For restaurants serving hot meals, add 10 W sensible and 10 W latent for food.

Table 1.3 Metabolic rate for different levels of activity

Activity	Metabolic rate	
	Met.	W/m²
Lying down	0.8	45
Seated quietly	1.0	58
Sedentary work – seated at work	1.2	70
Light activity – bodily movement on foot	1.6	93
Medium activity – bodily movement including carrying	2.0	117
High activity substantial physical work	3.0	175

heat sufficiently to its surroundings and may result in unconsciousness for the person. Hyperpyrexia can be induced internally through a fever or externally from high ambient temperature, high humidity and the effects of solar radiation incident upon the head in particular.

A person is likely to be comfortable, in the general sense of the word, when the heat generated by the body to maintain a core temperature of 37.2°C is equal to the heat lost to its surroundings.

The heat generated Q can be expressed as:

$$Q = +/- Q_c +/- Q_{cd} +/- Q_r + Q_e$$

See Figure 1.2. Heat conduction Q_{cd} takes place at points of physical contact. It constitutes a small proportion of the total and is usually ignored, thus:

$$Q = +/- Q_c +/- Q_r + Q_e$$

Sensible heat gain or loss therefore includes heat convection and heat radiation.

Latent heat loss from the body occurs at all times and provides evaporative cooling which includes insensitive perspiration from the skin surface, moisture evaporation from the process of breathing, and sweating. Latent heat loss from the body is therefore in three forms, namely

- passive moisture loss from the skin, which depends upon the vapour pressure of the surrounding air;
- moisture loss from the lungs, which depends upon ambient vapour pressure and breathing rate which in turn depends upon the degree of activity and thus the metabolic rate;
- active sweating, which commences when sensible heat loss plus insensible perspiration falls below the body's rate of heat production.

8 Thermal comfort and assessment

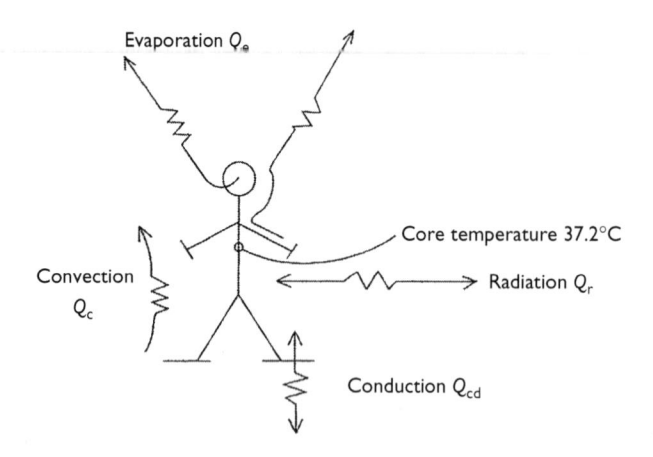

Figure 1.2 Latent heat loss and sensible heat loss/gain from the body.

Sweat is secreted by the eccrine glands which lie deep in the skin tissue. It consists of 99% water and 1% sodium chloride. The eccrine glands are activated by two control mechanisms:

Stimulus – peripheral receptors or sympathetic nerves;
Thermoregulation – the hypothalamus which is located in the brain and which responds to its own temperature variations.

The eccrine glands can be activated by heat energy resulting from physical work or the local climate or by physiological stimuli, especially for those glands located in the palms of the hands, the soles of the feet, the face and the chest.

For men working in the heat the sweat rate can reach 1 L/h. Hard work in a very hot environment may increase this rate to 2.5 L/h. This rate cannot be sustained for more than 30 min. Thus the bodily heat loss is restricted as the period of hard labour continues and core temperature is elevated. Refer to Table 1.2 for rates of body heat loss for different levels of activity. You will notice in this table that as the air temperature rises from 15°C to 26°C the proportions of latent to sensible heat change, but the total body heat loss for each level of activity does not change with air temperature rise. Thus for light work the total heat loss is 140 W and the ratio of latent to sensible heat loss at an air temperature of 15°C is $30/110 = 0.27$ whereas at an air temperature of 26°C the ratio is $70/70 = 1.0$. This represents a 3.7-fold increase in insensitive perspiration and sweat, assuming that the moisture evaporation from breathing is unchanged. You will be able to corroborate this evidence from

Thermal comfort and assessment 9

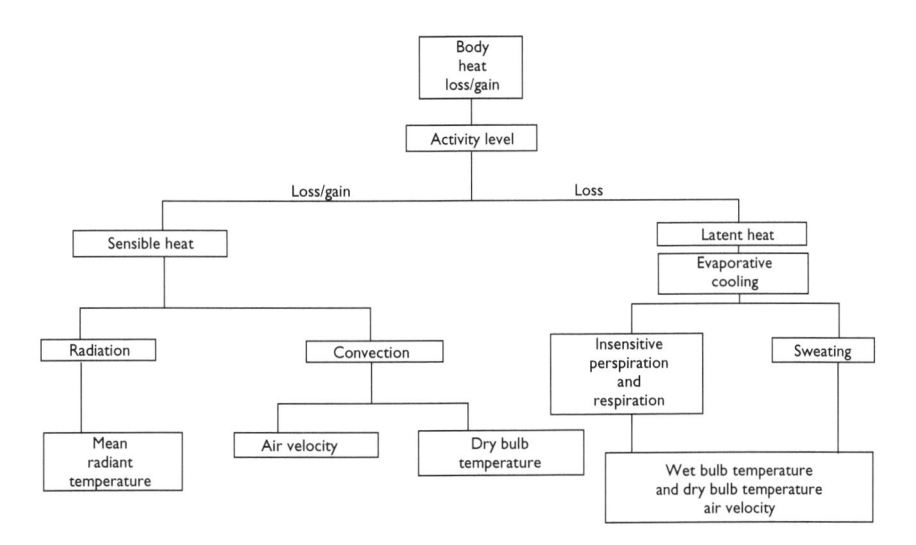

Figure 1.3 Heat energy flows between the human body and the surrounding climate.

personal experience of working or even sitting in surroundings of relatively high air temperature.

Figure 1.3 is a flow chart showing sensible and latent heat flows from the human body to the surrounding climate. Note the thermal criteria which trigger each mode of heat transfer. It was stated earlier that a heat balance can be drawn such that the heat generated to maintain a core temperature of 37.2°C is equal to the bodily heat loss to its surroundings. In thermally comfortable surroundings when mean radiant air and comfort temperature are say 20°C, a person doing light work absorbs approximately 24 L of oxygen per hour. For each litre of oxygen consumed, 21 kJ of heat is produced. Thus heat generated by the person $= 24 \times 21 \times 1000/3600 = 140$ W.

If the average temperature of exposed skin and clothing is 26°C and the surface area of the clothed body is 2 m^2, then, heat loss to the surroundings

$$Q = Q_c + Q_r + Q_e$$

Sensible heat loss.

$$Q_c = h_c \cdot A \cdot dt$$
$$Q_c = 3 \times 2 \times (26 - 20) = 36 \text{ W}$$
$$Q_r = eh_r \cdot A \cdot dt$$
$$Q_r = 5.7 \times 2 \times (26 - 20) = 68.4 \text{ W}$$

Latent heat loss. If the person loses 0.05 L/h and the latent heat of vaporisation is taken as 2500 kJ/kg

$$Q_e = (0.05/3600) \times 2500 \times 1000 = 34.7 \text{ W}$$

Thus, body heat loss

$$Q = 36 + 68.4 + 34.7$$
$$Q = 139 \text{ W}$$

and the heat generated by the body, calculated as 140 W, shows that the heat balance is maintained.

You should refer to Chapters 3 and 4 for the heat transfer coefficients h_c and eh_r for convection and radiation.

1.6 Physiological responses

Circulatory regulation of the blood flow is the initial response to thermal stress. In the subcutaneous layer which connects the skin to the surface muscles the regulation of blood flow is known as the vasomotor regulation. Regulation of the blood flow is achieved by vasodilation and vasoconstriction of the blood vessels. The vasomotor centre is located in the part of the brain known as the medulla oblongata.

The subcutaneous tissue has a high fat content and thus a high resistance to heat flow through the subcutaneous layer. Thus vasodilation within the subcutaneous layer induces large quantities of blood from the core through the subcutaneous tissue to the skin giving rise to high heat energy rejection. Vasoconstriction within the subcutaneous tissue induces low blood flow from the body core to the skin. These regulatory effects on the flow of blood vary the resistance to heat flow through the subcutaneous tissue. Thus the thermal resistance of the subcutaneous tissue which controls the heat flow at the skin surface is variable and responds to the degree of activity and ambient temperature in the manner described.

Low ambient temperature induces vasoconstriction in the subcutaneous tissue and hence the lower skin temperature for a person doing sedentary work. This inhibits excessive body heat loss to preserve core temperature. The sensation is the cooling of the extremities – fingers, nose, ears and toes.

Vasodilation is accompanied by an increase in heart rate, an increase in blood flow to the skin resulting in increased body heat loss and a reduced blood flow to the organs, and is induced by a high level of physical activity.

1.7 Thermal assessment

Human thermal comfort is a subjective condition which is witnessed by most of us fairly regularly: witness the varying amounts of clothing worn by different people in the same room. It is generally accepted, however, that there are four criteria which have a direct influence on human comfort:

- dry bulb temperature,
- wet bulb temperature,
- mean radiant temperature,
- air velocity.

Each of these criteria can be varied *within limits*, and still maintain comfort level, to compensate for one of them having a value outside the comfort range.

Many proposals for a thermal index which accounts for some or all of these criteria have been advanced in the last 100 years. The thermal indices in current use are dry resultant or comfort temperature and environmental temperature although the latter is not used now for the purposes of measurement.

- Air temperature is that which is measured by mercury in a glass thermometer shielded from direct heat radiation and suspended in air. The sensing bulb is small and as the mercury is reflective anyway, heat radiation incident on the bulb surface is insignificant, allowing the bulb to register local air temperature.
- Wet bulb temperature is obtained by placing a muslin sock over the sensing bulb of mercury in a glass thermometer and saturating it by placing the end of the sock in a container of distilled water. If the air local to the sensing bulb is dry, it has a low relative humidity and will evaporate the moisture in the sock. The rate of evaporation produces a cooling effect which will depress the mercury in the thermometer thus giving the wet bulb reading. The rate of evaporation on the sock of a wet bulb thermometer is proportional to the level of moisture in the local air, and a reading equal to the local dry bulb temperature implies that evaporation has ceased because the local air is saturated and relative humidity is, therefore, 100%.

The humidity range to ensure a satisfactory level of comfort is between 40% and 70%. The hand-held whirling sling psychrometer is still a popular instrument used for measuring dry bulb and wet bulb temperatures.

Atmospheric air

Air is a mixture of dry gases and water vapour where the water vapour is superheated and invisible to the eye. Thus at a room temperature of 20°C the gases which make up the dry air and the water vapour are at 20°C. Water at atmospheric pressure, however, boils at 100°C. This implies that for the vapour in the room to boil and be superheated at 20°C it must be at a pressure lower than the atmospheric pressure.

John Dalton (1766–1844) a British scientist developed a thesis that is called Dalton's Law of partial pressures that states:

> The total pressure of a gas mixture is equal to the sum of the partial pressures of the individual gases provided no chemical action occurs.

In the case of atmospheric air, the partial pressure exerted by the dry gases taken together and added to that of the water vapour make up atmospheric pressure.

Applying this law to air at 20°C db and 50.59% RH, from the *CIBSE Hygrometric Data*, the partial pressure of the water vapour is 11.82 mbar and dew point temperature is 9.4°C. Dew point temperature is the boiling point of water at these temperature and pressure conditions. It is the point at which condensation begins to form on the room surfaces that are at or below 9.4°C. As the dry air and the water vapour in the room are at 20°C, the amount by which the vapour is superheated will be $(20 - 9.4) = 10.6$ K. The partial pressure of the vapour is 11.82 mbar and taking atmospheric pressure as 101 mbar, the partial pressure of the dry air will therefore be $(101 - 11.82) = 89.18$ mbar.

Mean radiant temperature can be approximately evaluated from the mean weighted enclosure temperature, mwet.

$$\text{mwet} = (A_w \cdot t_w + A_f \cdot t_f + A_r \cdot t_r + A_g \cdot t_g)/(A_w + A_f + A_r + A_g)$$

thus

$$\text{mwet} = \Sigma(A \cdot t_s)/\Sigma A$$

The suffixes refer to the surface temperature of the enclosing walls, floor, roof and glazing.

You can see that it is the mean area weighted temperature of all the surfaces forming the enclosure. It is approximate because it is only most nearly a true mean radiant temperature if the point of measurement is in the centre of the space. Mean radiant temperature can be measured with the aid of a globe thermometer. See Figure 1.4. The matt finished sensing surface is greatly enlarged and is therefore more sensitive to absorb heat radiation than to sense the temperature of the air local to it. As air velocity increases, however, this instrument becomes less effective at measuring mean radiant temperature. See Example 1.1. The formula for globe temperature is:

$$t_g = (t_r + 2.35 t_a (u)^{0.5})/(1 + 2.35(u)^{0.5})$$

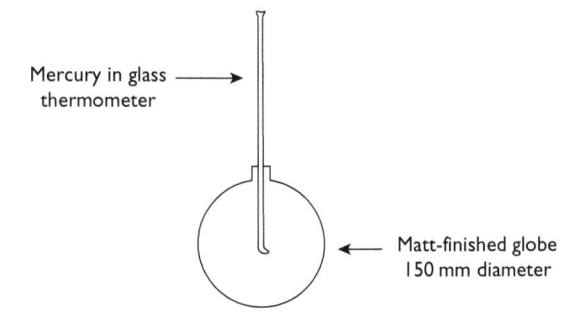

Figure 1.4 The globe thermometer.

from which

$$t_r = t_g(1 + 2.35(u)^{0.5}) - (2.35t_a(u)^{0.5})$$

When air velocity $u = 0.1$ m/s,

$$t_g = 0.57t_r + 0.43t_a \quad \text{and} \quad t_r = 1.75t_g - 0.75t_a$$

You should confirm that you agree with these three equations which originate from the equation for globe temperature t_g.

Comfort temperature is measured by a similar instrument like that for globe temperature except that the sensing surface is a 100 mm sphere. See Figure 1.4. It is also known as dry resultant temperature and was introduced by a French man in 1931. The formula for comfort temperature is

$$t_c = (t_r + t_a(10u)^{0.5})/(1 + (10u)^{0.5})$$

Another formula for t_c is given in the *CIBSE Guide*, section A 1970 edition as:

$$t_c = (t_r + 3.17t_a(u)^{0.5})/(1 + 3.17(u)^{0.5})$$

This formula is in the same format as that for globe temperature. Both the formulae will give the same results. When air velocity

$$u = 0.1 \text{ m/s},$$
$$t_c = 0.5t_r + 0.5t_a$$

Note: The above formulae for globe temperature and comfort temperature are empirical and therefore subject to some error.

14 Thermal comfort and assessment

Environmental temperature is not easily measured by an instrument and this is the main reason why it is not in common use now. For a cubical room in which local air velocity is around 0.1 m/s,

$$t_e = 0.667 t_r + 0.333 t_a$$

Low air speed is measured using a Kata thermometer. The time taken for warmed fluid to cool and contract down a glass stem between two fixed points is noted and use made of a nomogram provided by the manufacturer to convert the time to an air speed.

Note: Electronic instruments for measuring t_a, t_r, t_w, t_c and u are available.

You will see that comfort temperature and environmental temperature account for three of the four factors which influence thermal comfort. These are air temperature, mean radiant temperature and air velocity. Wet bulb temperature and hence relative humidity and vapour pressure is not accounted for and cannot be measured by a comfort temperature instrument. However, comfort temperature is currently the accepted thermal index in space heating design. A separate instrument is required to register relative humidity.

You will notice from the formula that at an air speed of 0.1 m/s which is an acceptable value in a room not subject to forced air movement, comfort temperature represents the sum of 50% air temperature and 50% mean radiant temperature. The formula for environmental temperature on the other hand is weighted towards mean radiant temperature.

For higher air speeds the equations for t_c and t_g change by varying the proportions of air and mean radiant temperature. Consider the following example.

Example 1.1 Determination of equations for comfort and mean radiant temperature

Determine the equations for comfort temperature and mean radiant temperature when air velocity is found to be 0.4 m/s.

Solution
It was shown earlier that:

$$t_g = (t_r + 2.35 t_a (u)^{0.5}) / (1 + 2.35 (u)^{0.5})$$

Substituting $u = 0.4$ m/s,

$$t_g = (t_r + 2.35 t_a (0.4)^{0.5}) / (1 + 2.35 (0.4)^{0.5})$$
$$= (t_r + 1.4863 t_a) / 2.4863$$
$$= (t_r / 2.4863) + (1.4863 t_a / 2.4863)$$

Thermal comfort and assessment

from which

$$t_g = 0.4t_r + 0.6t_a$$
$$t_r = (t_g - 0.6t_a)/0.4$$

from which $t_r = 2.5t_g - 1.5t_a$.

Now, from earlier in this section it was shown that:

$$t_c = (t_r + t_a(10u)^{0.5})/(1 + (10u)^{0.5})$$

substituting $u = 0.4$ m/s,

$$t_c = (t_r + 2t_a)/3$$

from which $t_c = 0.33t_r + 0.67t_a$.

It is now appropriate to analyse the effects that air temperature and globe temperature have upon mean radiant temperature and comfort temperature at different air velocities. Consider the following case study.

Case study 1.1 Evaluation of comfort and mean radiant temperatures for sedentary occupation

A heated room is used for sedentary occupation.

(a) Evaluate comfort and mean radiant temperature for the room in which the measured globe and air temperatures are 17°C and 21°C, respectively for air velocities of 0.1 m/s and 0.4 m/s.
(b) Evaluate comfort and mean radiant temperature for the room in which the measured globe and air temperatures are 21°C and 17°C, respectively for air velocities of 0.1 m/s and 0.4 m/s.
(c) Summarise and draw conclusions from the results.

Solution

(a) When $t_g = 17°C$, $t_a = 21°C$ and $u = 0.1$ m/s,

$$t_r = 1.75t_g - 0.75t_a = 1.75 \times 17 + 0.75 \times 21 = 14°C$$
$$t_c = 0.5t_r + 0.5t_a = 0.5 \times 14 + 0.5 \times 21 = 17.5°C$$

When $u = 0.4$ m/s,

$$t_r = 2.5t_g - 1.5t_a = 2.5 \times 17 - 1.5 \times 21 = 11°C$$
$$t_c = 0.33t_r + 0.67t_a = 0.33 \times 11 + 0.67 \times 21 = 17.67°C$$

(b) When $t_g = 21°C$, $t_a = 17°C$ and $u = 0.1$ m/s

$$t_r = 1.75 \times 21 - 0.75 \times 17 = 24°C$$
$$t_c = 0.5 \times 24 + 0.5 \times 17 = 20.5°C$$

When $u = 0.4$ m/s,

$$t_r = 2.5 \times 21 - 1.5 \times 17 = 27°C$$
$$t_c = 0.33 \times 27 + 0.67 \times 17 = 20.3°C$$

(c) The analysis is summarised in Table 1.4.

CONCLUSIONS FROM THE SUMMARY OF CASE STUDY 1.1

There are two comfort zones which can be applied here.

- For mainly sedentary occupations, comfort temperature t_c should fall between 19°C and 23°C. Cases 1 and 2 therefore fall outside this comfort zone and may not be conducive to thermal comfort.
- The difference in temperature between mean radiant and air temperature should be within an envelope of +8 K or −5 K for mainly sedentary occupations. Cases 1, 2 and 4 therefore fall outside this comfort zone and may not provide a satisfactory level of thermal comfort. Case 3 thus appears to be the only right solution here for sedentary occupation.

Current standards of thermal insulation recommended in the Building Regulations along with limiting the infiltration rate of outdoor air will mitigate in favour of keeping the difference between mean radiant temperature and air temperature within the +8 K, −5 K envelope. (This matter is discussed in detail in another publication in the series.)

- The ranking of the thermal indices in the heat flow paths is dependent upon the type of space heating. Cases 3 and 4 indicate a mainly radiant system of space heaters. Cases 1 and 2 indicate a mainly convective heating system.

Table 1.4 Summary of Case study 1.1

Measured u (m/s)	Measured t_g	Measured t_r	Calculated t_a	Calculated t_c	Case	Heat flow paths	$(t_r - t_a)$ Temperature difference (K)
0.1	17	14	21	17.5	1	$t_a \to t_c \to t_g \to t_r$	−7
0.4	17	11	21	17.67	2	$t_a \to t_c \to t_g \to t_r$	−10
0.1	21	24	17	20.5	3	$t_r \to t_g \to t_c \to t_a$	7
0.4	21	27	17	20.3	4	$t_r \to t_g \to t_c \to t_a$	10

- The most acceptable level of comfort for sedentary occupations is achieved when the difference between mean radiant and air temperature is within a +8 K, −5 K envelope and the comfort temperature zone of between 19°C and 23°C, with relative humidity between 40% and 70%.
- If electronic temperature measurement equipment is used you can see the importance of establishing which thermal index the instrument is actually measuring.

1.8 Thermal comfort criteria

Professor O. Fanger has spent much time both in Denmark and the United States researching how thermal comfort can be assessed for a variety of occupations. His work has now been encompassed in BSEN ISO 7730/1995. The metabolic rate can vary from 30 to 500 W/m² depending upon the level of activity. One 'Met.' is equal to the metabolic rate for a seated person and is equivalent to 58 W/m².

The thermal resistance of clothing is measured in units of 'Clo' such that:

1 Clo unit = 0.155 m²K/W

Tog is the European unit of thermal resistance for bedding and

1 Tog = 0.645 Clo = 0.1 m²K/W

Table 1.5 gives typical insulation levels for some clothing ensembles.

In his research on thermal assessment Professor Fanger uses the term ambient temperature (which is often used as a term for outdoor temperature) to refer to indoor conditions when air and mean radiant temperature are the same in value. The comfort diagrams which he produced can be used by HVAC engineers as well as those practising occupational hygiene and by health and safety officers. Three of the diagrams are shown in Figures 1.5, 1.6 and 1.7 and the following examples will demonstrate their use.

Table 1.5 Clothing insulation levels – the relationship between the Clo unit and thermal conductivity R

Clothing combination	Insulation level	
	Clo	R (m²K/W)
Naked	0	0
Shorts/bikini	0.1	0.016
Light summer clothing	0.5	0.078
Indoor winter clothing	1.0	0.155
Heavy suit	1.5	0.233
Polar weather suit	3–4	0.465–0.62
Tog	0.645	0.1

18 Thermal comfort and assessment

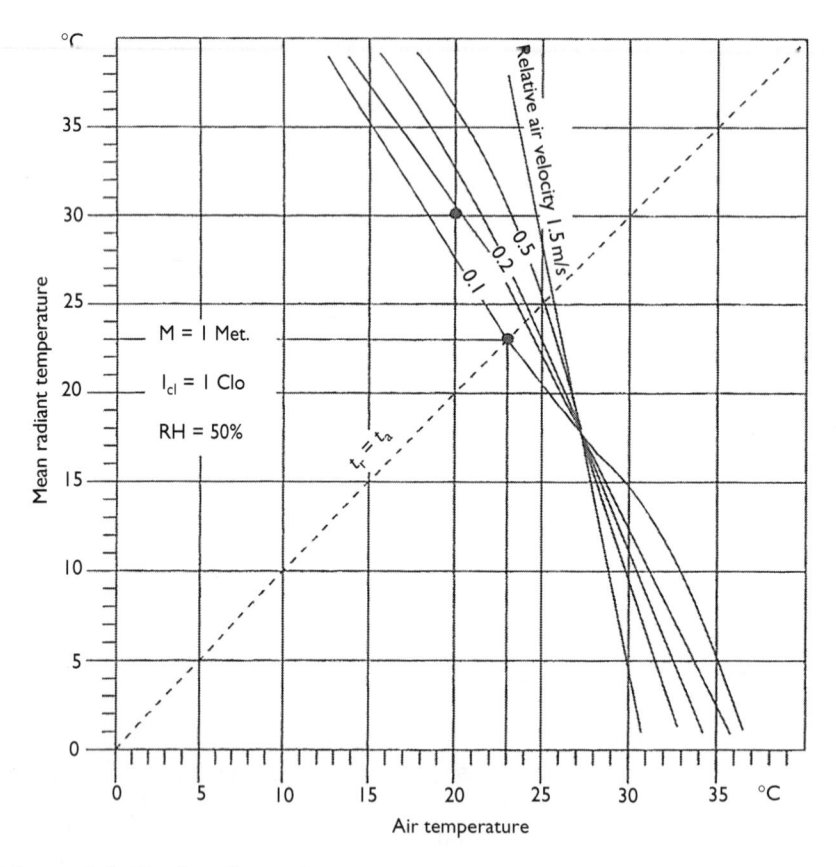

Figure 1.5 Comfort diagram 1.

Example 1.2 Determination of required ambient temperature for an office

The staff occupying an office are clothed at 1 Clo and are undertaking sedentary activities. Determine the required ambient temperature when air velocity is 0.1 m/s.

Solution
Ambient temperature $t_{am} = t_r = t_a$. Using Figure 1.5, points on the dotted diagonal line yield equal mean radiant and air temperature. Where the dotted line intersects the isovel (line of constant velocity) of 0.1 m/s, $t_{am} = 23°C$.

It is not usually possible to attain equal mean radiant and air temperatures which are dependent upon the level of insulation of the building, the ventilation rate and the type of heating system. They should, however, be close to this ambient temperature.

Thermal comfort and assessment 19

Example 1.3 Determination of mean radiant temperature for an occupied store room

A store room is held at an air temperature of 16°C and 50% relative humidity. The air velocity is 0.1 m/s. A storeman wearing clothing of 1 Clo is allocated to work in the room. If his activity level is 1 Met. find the mean radiant temperature required to provide the right level of comfort.

Solution

From Table 1.3, 1 Met. represents a person seated quietly. It is likely therefore that the storeman is located at a desk or a counter.

From the comfort diagram in Figure 1.5, mean radiant temperature $t_r = 33°C$. This is well outside the second comfort zone used in Case study 1.1. However, the whole store does not need radiant heating to this level. A luminous directional radiant heater located at low level would be recommended. It would also be worth suggesting an alternative clothing combination at least to 1.5 Clo. See Table 1.5.

Example 1.4 Evaluate the conditions that provide thermal comfort for spectators

Determine the conditions that will provide thermal comfort for seated spectators at a sports centre swimming pool. Their clothing combination is 0.5 Clo and their activity level is 1 Met. Assume that the air velocity is 0.2m/s and relative humidity is 80%.

Solution

From Table 1.3, 1 Met. refers to a person seated quietly and from Table 1.5, 0.5 Clo is equivalent to light summer clothing. From the comfort diagram in Figure 1.6, air and mean radiant temperature should be 26.2°C.

Example 1.5 Finding the ambient temperature required for store assistants

Assistants working in a retail store have an estimated activity level of 2 Met. The clothing worn has an insulation level of 1 Clo. If the air velocity is 0.4 m/s, find the ambient temperature which should provide thermal comfort for the assistants.

Solution

From Table 1.3, 2 Met. indicates medium activity and from Table 1.5, 1 Clo relates to indoor winter clothing. From the comfort diagram in Figure 1.7, air and mean radiant temperature should be 17.2°C and relative humidity 50%.

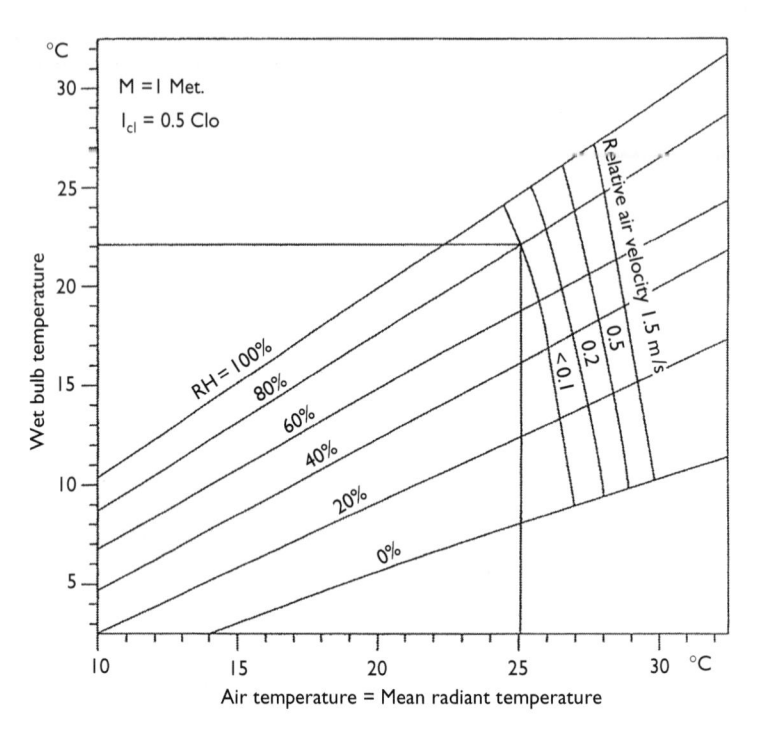

Figure 1.6 Comfort diagram 2.

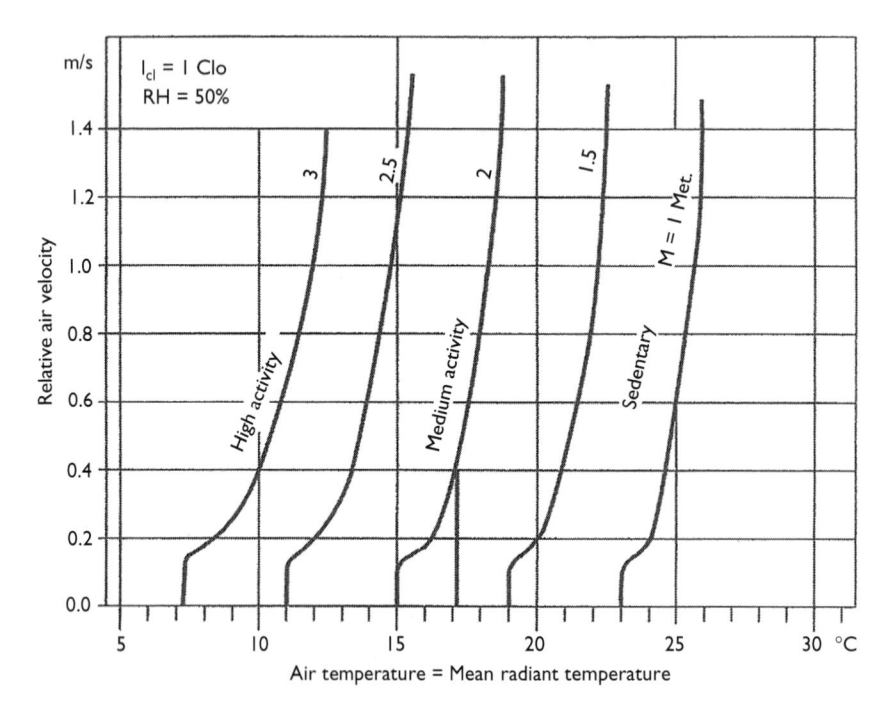

Figure 1.7 Comfort diagram 3.

Example 1.6 Changing air temperature controls to maintain thermal comfort

An office is staffed by personnel whose clothing insulation value is 1 Clo and who are engaged in activity estimated at 1 Met. Relative humidity is 50%, air velocity 0.1 m/s and mean radiant and air temperature is 23°C. Installation of new business equipment increases the mean radiant temperature to 26°C. By what amount should the air temperature controls be changed to maintain thermal comfort?

Solution

From the comfort diagram in Figure 1.5, for a mean radiant temperature of 26°C, required air temperature is 21°C. This will mean re-setting the controls $(23 - 21) = 2$ K lower than before. It is likely that by doing this the mean radiant temperature of 26°C will be lowered and therefore fine tuning may be necessary.

Example 1.7 Recommending ambient conditions for factory operatives

Factory operatives work in an environment in which the air velocity is 0.5 m/s and relative humidity is 50%. The factory overalls provide an insulation value of 1 Clo and the activity level is expected to be 2 Met. Recommend suitable ambient conditions for the factory.

Solution

From the comfort diagram in Figure 1.7, air and mean radiant temperature should be 17.5°C.

Qualifying satisfaction

ASHRAE has developed a seven-point scale of assessment for thermal environments and Professor Fanger has developed a method of predicting the level of satisfaction using the predicted mean vote (PMV). Table 1.6 gives the ASHRAE scale and the corresponding PMV.

Professor Fanger converted the PMV index to the predicted percentage of dissatisfied (PPD) people in order to show the PMV as a percentage of people occupying a thermal environment who would be likely to be dissatisfied with the level of comfort.

Figure 1.8, which is a normal distribution, shows the conversion from PMV to PPD. If, for example, the PMV index is -1 or $+1$, the predicted percentage dissatisfied, from Figure 1.8, is 26%. The other 74% are likely to be satisfied. When the PMV is zero (Table 1.6) the PPD, from Figure 1.8, is 5%. This implies, for example, that out of 40 people occupying a room in which the PMV is zero, 2 people are likely to be dissatisfied. The other 95% or 38 people will be satisfied.

22 Thermal comfort and assessment

Table 1.6 Comfort scales

ASHRAE scale	Fanger PMV index
Hot	3
Warm	2
Slightly warm	1
Neutral	0
Slightly cool	−1
Cool	−2
Cold	−3

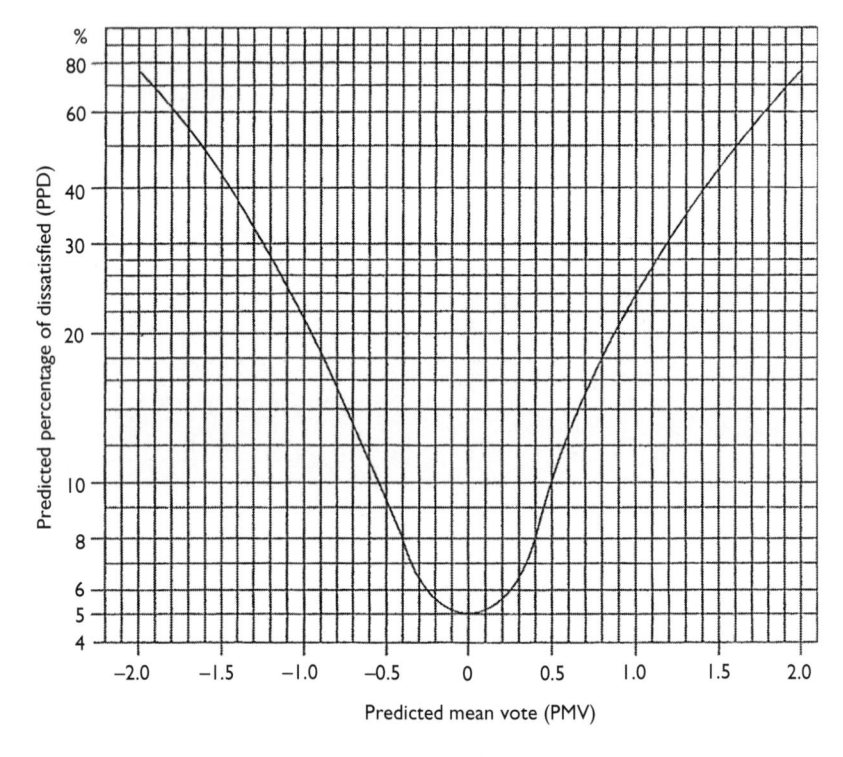

Figure 1.8 Predicted percentage of dissatisfied (PPD) as a function of predicted mean vote (PMV).

Quantifying the level of dissatisfaction

It is clearly important to ensure that the level of dissatisfaction with the thermal environment is kept to a minimum, preferably to no more than 5% of the people populating a building.

Table 1.7 allows the PMV index to be obtained for ambient temperatures between 20°C and 27°C and air velocities from < 1.0 to 1.0 m/s, when the relative humidity is 50%, the metabolic rate is 1 Met. and the clothing insulation level is 1 Clo. Table 1.8 gives the rate of change of PMV with ambient temperature [d(PMV)/d$_{tam}$] and Table 1.9 gives the rate of change of PMV with mean radiant temperature [d(PMV)/dt$_r$].

The four variables to be measured are: dry bulb temperature, relative humidity (wet bulb temperature), mean radiant temperature and air velocity.

Table 1.7 Predicted mean vote index

Ambient temperature (t°_{amb}C)	RH = 50%, M = 1 Met., I_{Clo} = 1 Clo Relative air velocity (U_a) (m/s)							
	<0.1	0.1	0.15	0.20	0.30	0.40	0.5	1.00
20	−0.85	−0.87	−1.02	−1.13	−1.29	−1.41	−1.51	−1.81
21	−0.57	−0.60	−0.74	−0.84	−0.99	−1.11	−1.19	−1.47
22	−0.30	−0.33	−0.46	−0.55	−0.69	−0.80	−0.88	−1.13
23	−0.02	−0.07	−0.18	−0.27	−0.39	−0.49	−0.56	−0.79
24	0.26	0.20	0.10	0.02	−0.09	−0.18	−0.25	0.46
25	0.53	0.48	0.38	0.13	0.21	0.13	0.07	−0.12
26	0.81	0.75	0.66	0.60	0.51	0.44	0.39	0.22
27	1.08	1.02	0.95	0.89	0.81	0.75	0.71	0.56

Table 1.8 Rate of change of PMV with ambient temperature (dPMV/dt$_{am}$)

Clothing insulation (I_{cl}) Clo	RH = 50%, M = 1 Met. Relative air velocity (U_a) (m/s)			
	<0.1	0.2	0.5	1.0
0.5	0.350	0.370	0.435	0.490
1.0	0.258	0.260	0.290	0.310
1.5	0.198	0.200	0.217	0.230

Table 1.9 Rate of change of PMV with mean radiant temperature (dPMV/dt$_r$)

Clothing insulation (I_{cl}) Clo	RH = 50%, M = 1 Met. Relative air velocity (U_a) (m/s)			
	<0.1	0.2	0.5	1.0
0.5	0.160	0.155	0.140	0.12
1.0	0.117	0.107	0.090	0.077
1.5	0.090	0.080	0.067	0.055

24 Thermal comfort and assessment

The point of measurement is normally taken as the centre in plan and the height above floor is normally 0.6 m for sedentary occupancy and 1.0 m for standing occupancy.

Example 1.8 Determination of the predicted percentage dissatisfied (PPD)

A factory has the thermal conditions given below. It is proposed to section off part of the factory for employees clothed at 1 Clo and engaged in sedentary work at 1 Met. Determine the PPD for the area and determine the value to which the mean radiant temperature must be raised to provide an acceptable level of thermal comfort.

Data: Air temperature 21°C, mean radiant temperature 16°C, air velocity 0.2 m/s, relative humidity 50%.

Solution

From reference to the thermal comfort diagram in Figure 1.5, the intersection of the air and mean radiant temperatures are to the left of the 0.2 m/s velocity isovel. The conditions therefore are too cool and do not produce thermal comfort. From Table 1.8, ambient temperature is taken as air temperature of 21°C and at an air velocity of 0.2 m/s, $PMV_{ta} = -0.84$. Since mean radiant temperature is too low, $[d(PMV)/dt_r]$ from Table 1.9 is obtained where air velocity is 0.2 m/s and clothing insulation is 1 Clo. Thus $[d(PMV)/dt_r] = 0.107$. The PMV correction factor,

$$PMV_c = [d(PMV)/dt_r](t_a - t_r) = 0.107(21 - 16) = 0.535$$

and $PMV = PMV_{ta} + PMV_c = -0.84 - 0.535 = -1.375$. From Figure 1.8, PPD = 44%.

From the comfort diagram in Figure 1.5, given air temperature as 21°C and air velocity as 0.2 m/s, mean radiant temperature is read off as 29°C.

You should now undertake the solution based upon an air velocity of 0.1 m/s.

1.9 Temperature profiles

The vertical temperature profile in the conditioned space will vary with the type of heating/cooling employed to offset the heat losses/gains. Figure 1.9 shows temperature profiles resulting from different types of space heating. The ideal profile is one which is close to vertical. Low air and mean radiant temperatures at floor level or high temperatures at head level will encourage levels of discomfort.

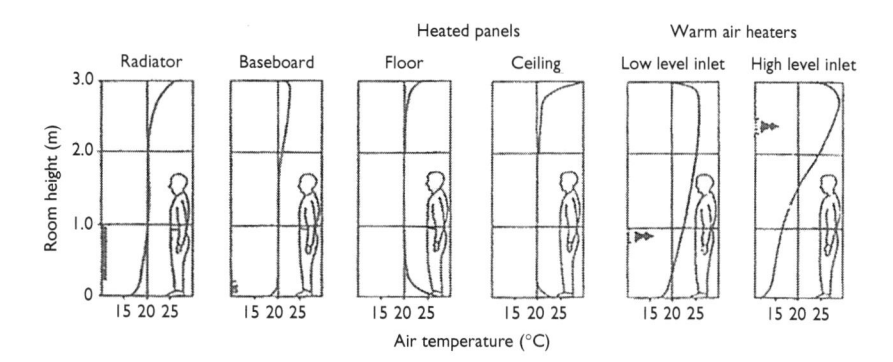

Figure 1.9 Vertical temperature profiles.

Source: Reproduced from the *CIBSE Guide* (1986) by permission of the Chartered Institution of Building Services Engineers.

1.10 Chapter closure

This chapter has provided you with the underpinning knowledge of the response mechanisms of the human body to surrounding air and mean radiant temperatures and of the thermal comfort criteria which have a direct bearing upon the level of comfort. The thermal indices used to measure these criteria have been identified and used to analyse comfort levels. This should encourage you to pursue, investigate and analyse other situations. Personal thermal comfort is of course a subjective matter and therefore it is unlikely that all the occupants of a conditioned building will agree on the level of comfort provided within it. This should not act as a discouragement from attempts to ensure a thermally comfortable environment indoors for the majority of occupants.

Chapter 2

Heat conduction

Nomenclature

A	surface area (m^2)
A_i	inner surface area (m^2)
A_m	logarithmic mean surface area (m^2)
A_o	outer surface area (m^2)
dt	temperature difference (K)
e	emissivity of surface
EPS	expanded polystyrene
h_c, eh_r	heat transfer coefficient for convection/radiation (W/m^{2}K)
h_s	surface conductance (W/m^{2}K), taken as $(h_c + eh_r)$
h_{si}, h_{so}	surface conductance at the inner/outer surface (W/m^{2}K)
I	heat flux (W/m^2)
k	thermal conductivity (W/mK)
L	thickness (m)
Q	heat flow (W)
Q/L	heat flow per unit length (W/m)
R, R_e, R_g	thermal resistance (m^{2}K/W)
R_a, R_v	thermal resistance of the air cavity/void (m^{2}K/W)
R_c	reciprocal of h_c
R_r	reciprocal of eh_r
R_{si}, R_{so}	thermal resistance at the inside/outside surface (m^{2}K/W)
R_t	total thermal resistance (m^{2}K/W)
R_{t1}, R_{t2}	transform thermal resistances (m^{2}K/W)
t_{ai}	indoor air temperature (°C)
t_{ao}	outdoor air temperature (°C)
t_c	dry resultant temperature (°C)
t_{ei}	indoor environmental temperature (°C)
t_{eo}	outdoor environmental temperature (°C)
t_i	indoor temperature (°C)
t_o	outdoor temperature (°C)
t_r	mean radiant temperature (°C)
t_s	surface temperature (°C)
U	thermal transmittance coefficient (W/m^{2}K)

2.1 Introduction

Heat conduction can occur in solids. It can also occur in liquids and gases in which the vibrating molecules are unable to break free from each other because of the presence of boundary surfaces having a small temperature differential.

The air gap in an external cavity wall of a heated building at normal temperatures will contain still air since the temperature difference between indoors and outdoors is insufficient for the air to generate convection currents resulting from the density difference of the air between the inside surface and the outside surface of the cavity. If the air gap is increased to more than 75 mm, there is sufficient room for the air in the cavity to overcome its own viscosity and the resistance at the inside surfaces of the cavity, and the difference in density will encourage the air to convect naturally. The thermal conductivity k of still air is around 0.024 W/mK making it an excellent thermal insulation. Convected air on the other hand provides very poor thermal insulation.

Heat conduction also takes place in stationary films of fluids which occurs at boundary surfaces such as in air at the inner and outer surface of the building envelope and on the inside surface of pipes and ducts in which water and air respectively are flowing. In these circumstances, the heat conduction is considered either as a surface conductance h_{si}, h_{so} or as a surface thermal resistance R_{si} and R_{so}. The *CIBSE Guide* lists typical values of thermal conductivity for different materials employed in the building process. This is condensed in Table 2.1.

Thermal conductivity is one of the properties of a substance. In liquids and gases it is affected by changes in temperature more than in solids. The thermal conductivity of porous solids is affected by the presence of moisture. This is the reason why in Table 2.1 the outer leaf of a brick wall has a higher thermal conductivity than the inner leaf. Heat conduction can be considered as taking place radially outwards as in the case of an insulated pipe transporting a hot fluid; in two directions as in the case of a floor in contact with the ground and air; and in one direction as in the case of heat flow at right angles through the external building envelope.

2.2 Heat conduction at right angles to the surface

This mode of heat transfer is commonly associated with that through the building structure. See Figure 2.1. Fourier's law for one-dimensional steady-state heat flow through a single slab of homogeneous material at right angles to the surface is:

$$I = k \cdot dt/L \text{ W/m}^2$$

where dt = temperature difference across the faces of the slab.

Table 2.1 Properties of materials used in buildings

Material	Density (kg/m³)	Thermal conductivity (W/m K)	Specific heat capacity (J/kg K)
Walls (External and Internal)			
Asbestos cement sheet	700	0.36	1050
Asbestos cement decking	1500	0.36	1050
Brickwork (outer leaf)	1700	0.84	800
Brickwork (inner leaf)	1700	0.62	800
Cast concrete (dense)	2100	1.40	840
Cast concrete (lightweight)	1200	0.38	1000
Concrete block (heavyweight)	2300	1.63	1000
Concrete block (mediumweight)	1400	0.51	1000
Concrete block (lightweight)	600	0.19	1000
Fibreboard	300	0.06	1000
Plasterboard	950	0.16	840
Title hanging	1900	0.84	800
Surface finishes			
External rendering	1300	0.50	1000
Plaster (dense)	1300	0.50	1000
Plaster (lightweight)	600	0.16	1000
Roofs			
Aerated concrete slab	500	0.16	840
Asphalt	1700	0.50	1000
Felt/bitumen layers	1700	0.50	1000
Screed	1200	0.41	840
Stone chippings	1800	0.96	1000
Tile	1900	0.84	800
Wood wool slab	500	0.10	1000
Floors			
Cast concrete	2000	1.13	1000
Metal trap	7800	50.00	480
Screed	1200	0.41	840
Timber flooring	650	0.14	1200
Wood blocks	650	0.14	1200
Insulation			
Expanded polystyrene (EPS) slab	25	0.035	1400
Glass fibre quilt	12	0.040	840
Glass fibre slab	25	0.035	1000
Mineral fibre slab	30	0.035	1000
Phenolic foam	30	0.040	1400
Urea formaldehyde (UF) foam	10	0.040	1400

Source: Reproduced from the *CIBSE Guide* (1986) by permission of the Chartered Institution of Building Services Engineers.

Notes
Surface resistances have been assumed as follows
External walls $R_{so} = 0.06$ m² K/W
  $R_{si} = 0.12$ m² K/W
  $R_a = 0.18$ m² K/W
Roofs $R_{so} = 0.04$ m² K/W
  $R_{si} = 0.10$ m² K/W
  $R_{so} = 0.18$ m² K/W (pitched)
  $R_a = 0.16$ m² K/W (flat)
Internal walls $R_{so} = R_{si} = 0.12$ m² K/W
  $R_a = 0.18$ m² K/W
Internal floors $R_{so} = R_{si} = 0.12$ m² K/W
  $R_a = 0.20$ m² K/W

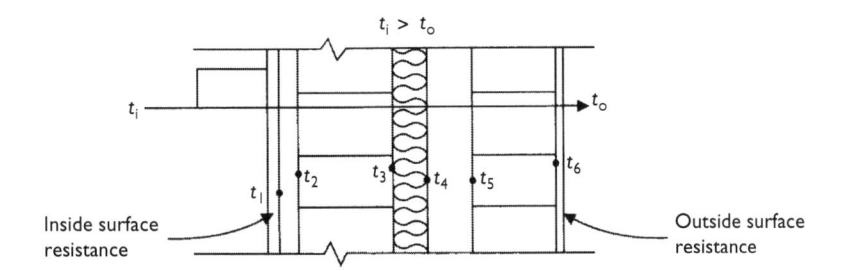

Figure 2.1 Conductive heat flow through a composite wall.

Source: Reproduced from the *CIBSE Guide* (1986) by permission of the Chartered Institute of Building Services Engineers.

This law has its limitations because we usually work from the temperature indoors to the temperature outdoors and this includes the inside and outside surface resistances R_{si} and R_{so}. Furthermore, there is frequently more than one slab of material in the building structure being considered.

A more appropriate generic formula is given below in which indoor temperature t_r, t_{ai}, t_{ei} and t_c are for convenience considered to be equal in value and denoted here as t_i. Outdoor temperature t_{ao} and t_{eo} are considered equal and are denoted here as t_o.

$$R_t = (1/h_{si}) + \sum(L/k) + R_a + 1/h_{so} = 1/U \text{ m}^2\text{K/W}$$

where the thermal resistance R for each element in the composite structure is

$$R = L/k \text{ m}^2\text{K/W}$$

In the context of heat conduction the reciprocal of the surface conductance h_{si}, h_{so} at the inside and outside surfaces is normally used thus:

$$R_{si} = 1/h_{si} \quad \text{and} \quad R_{so} = 1/h_{so}$$

and the generic formula becomes:

$$R_t = R_{si} + \sum(L/k) + R_a + R_{so} = 1/U \text{ m}^2\text{K/W} \tag{2.1}$$

from which the thermal transmittance coefficient or U value for the composite structure which includes surface film resistances is calculated. It follows that the intensity of heat flow (heat flux) I will be:

$$I = U \cdot (t_i - t_o) \text{ W/m}^2 \tag{2.2}$$

30 Heat conduction

Furthermore, from equations 2.1 and 2.2

$$I = dt/R_t \text{ W/m}^2$$

If indoor to outdoor temperature difference dt is steady the heat flux I will be steady. Thus, $dt/R_t = I =$ constant and therefore,

$$dt \propto R$$

So

$$dt_1/R_1 = dt_2/R_2 \text{ W/m}^2 \qquad (2.3)$$

This allows the determination of face and interface temperatures in a composite structure at steady temperatures.

The conductive heat flow through the composite structure may be determined from:

$$Q_s = U \cdot A \cdot (t_i - t_o) \text{ W}$$

Consider the conductive heat flow path through a composite structure having two structural elements and an air cavity, Figure 2.2. Equation 2.3 may be adapted as follows:

$$(t_i - t_1)/R_{si} = (t_i - t_o)/R_t$$

from which t_1 can be determined. Similarly:

$$(t_i - t_2)/(R_{si} + R_1) = (t_i - t_o)/R_t$$

from which t_2 can be calculated, and so on.

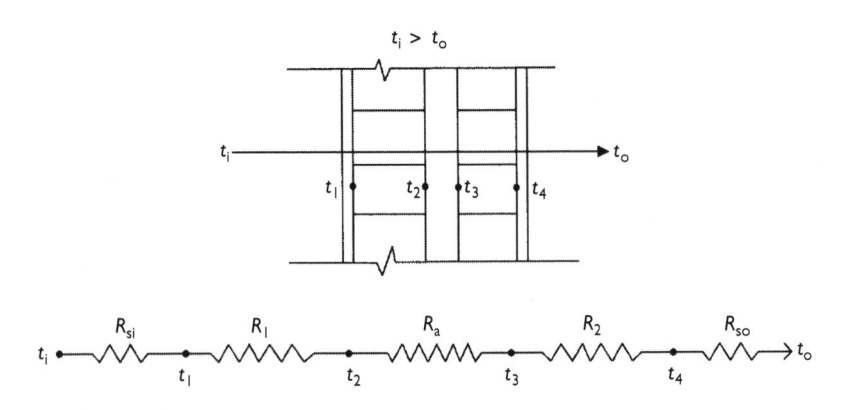

Figure 2.2 Heat flow path through a composite structure.

Radiant heating

Convective heating

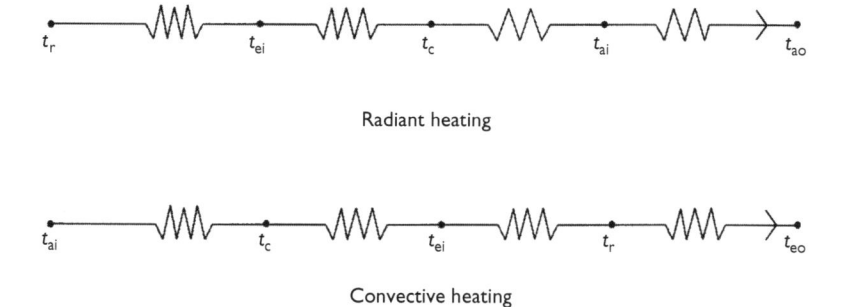

Figure 2.3 Heat flow paths.

The conductive heat flow through a composite structure may be adapted to determine the heat flow, through an external building envelope, for example, consisting of a number of different structures, such as walls, glazing, floor, roof and doors as follows:

$$Q_s = \sum (UA)(t_i - t_o) \text{ W} \qquad (2.4)$$

It is not the purpose of this book to undertake building heat loss calculations which must include also the heat loss due to infiltration and which are conveniently done using appropriate software, the manual method being analysed in another publication.

However, it is appropriate here to point out that in practice the thermal indices t_r, t_c, t_{ei} and t_{ai} are rarely equal in a heated building. This is due to the type of heating system adopted and to an extent upon the level of thermal insulation and natural ventilation. For example, the heat flow path for a building heated by radiant heaters is different to that heated by fan coil units or unit heaters. In the first case, the radiant component of heat transfer may be as high as 90% of the total, whereas in the second case the convective component of heat transfer will be 100% with no component of heat radiation at all. This clearly has an effect on the heat flow paths through the thermal indices. See Figure 2.3. Equation 2.4 therefore will provide an approximate structural heat loss. A more accurate methodology involves the surface conductance h_s.

2.3 Surface conductance

Heat conductance h_s in the surface film combines the coefficients of heat transfer for convection h_c and radiation eh_r thus:

$$h_s = eh_r + h_c \text{ W/m}^2\text{K}$$

32 Heat conduction

For indoor temperatures around 20°C and using an average value for the heat transfer coefficient for convection and a typical emissivity with the radiation coefficient:

$$h_s = (0.9 \times 5.7) + 3.0 = 8.13 \text{ W/m}^2\text{K}$$

from which $R_{si} = 1/h_s = 1/8.13 = 0.123 \text{ m}^2 \text{ K/W}$. This agrees with the value for R_{si} listed in Table 2.1 and assumes that t_r and t_{ai} are equal. Since this rarely is the case, the resulting conductive heat flow must be approximate but is considered good enough for most U value calculations.

A more accurate calculation can be obtained by separating the components of convection and radiation thus:

$$I = eh_r(t_r - t_s) + h_c(t_{ai} - t_s) \text{ W/m}^2 \tag{2.5}$$

Example 2.1 Conductive heat flow through an external wall

A room is held at an air temperature of 20°C. When outdoor temperature is −1°C and under steady conditions the inside surface temperature of the external wall is measured at 17.5°C. Assuming a wall emissivity of 0.9, determine

(i) the rate of conductive heat flow through the wall if $t_r = t_s$;
(ii) the rate of heat flow through the wall when $t_r = t_{ai}$;
(iii) the rate of heat flow through the wall when $t_r = 21°C$.

Solution

(i) Heat flux $I = eh_r(t_r - t_s) + h_c(t_{ai} - t_s)$

$$I = 3(20 - 17.5) = 7.5 \text{ W/m}^2 \text{ of wall.}$$

(ii) $I = 0.9 \times 5.7(20 - 17.5) + 3(20 - 17.5)$

$$I = 20.325 \text{ W/m}^2 \text{ of wall.}$$

Also, $I = h_s((t_r, t_{ai}) - t_s) = 8.13(20 - 17.5) = 20.325 \text{ W/m}^2$, or $I = ((t_r, t_{ai}) - t_s)/R_{si} = (20 - 17.5)/0.123 = 20.325 \text{ W/m}^2$.

(iii) Clearly in this case the room is radiantly heated for $t_r > t_s$ and

$$I = 0.9 \times 5.7(21 - 17.5) + 3(20 - 17.5)$$

$$I = 25.455 \text{ W/m}^2 \text{ of wall.}$$

SUMMARY FOR EXAMPLE 2.1
Note the variations in conductive heat flow in the table given below.

Case	t_{ai}	t_r	t_s	t_o	Heat flux (W/m²)
(i)	20	17.5	17.5	−1	7.5
(ii)	20	20	17.5	−1	20.325
(iii)	20	21	17.5	−1	25.455

In case (ii) we have the standard method for determining the U value which is when t_r is assumed to be the same as t_{ai}, thus from equation 2.2, $I = U \cdot dt$ and therefore the external wall $U = I/dt = 20.325/(20+1) = 0.968$ W/m²K.

Clearly the heat flux in case (iii) is greater than in case (ii). This will have the effect of increasing the structural heat loss in radiantly heated buildings. The determination of plant energy output Q_p (which is discussed in detail in another publication in the series) accounts for this, whereas the calculation of structural heat loss adopting equation 2.4 may not and may therefore only provide an approximation of structural heat flow at steady temperatures as the summary to Example 2.1 shows. The heat flow paths for the wall are shown in Figure 2.4. Appendix 2, Example A2.7 analyses the time it takes to raise the inner surface temperature of an external wall subjected to a calculated heat flux I W/m².

Example 2.2 Finding the conductive heat flux and interface temperatures through an external wall

An external cavity wall is constructed from the components shown below. From the data, determine the rate of conductive heat flux at right angles to the surface and calculate the temperature at each face and interface.

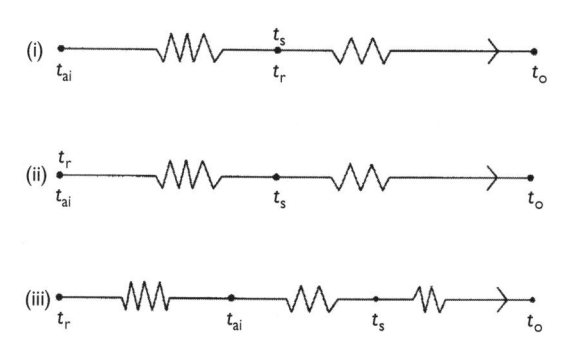

Figure 2.4 Example 2.1 – heat flow paths.

34 Heat conduction

Wall construction: Inner leaf 10 mm lightweight plaster, 110 mm lightweight concrete block lined on the cavity face with 25 mm of glass fibre slab, air space 50 mm, outer leaf 110 mm brick. See Figure 2.1.

Data: Indoor air temperature 23°C, indoor mean radiant temperature 18°C, outdoor temperature −2°C.

Thermal conductivities and outside surface resistance are taken from Table 2.1. Inside surface convective heat transfer coefficient, $h_c = 3.0$ W/m²K. Inside surface heat transfer coefficient for radiation, $eh_r = 5.13$ W/m²K.

Solution

If temperatures remain steady, a heat balance may be drawn such that: heat flow from the indoor radiant and air points to the inside surface t_s = heat flow from the inside surface t_s to the outdoor temperature t_o. Combining equations 2.3 and 2.5 the heat balance becomes:

$$eh_r(t_r - t_s) + h_c(t_{ai} - t_s) = (t_s - t_o)/R$$

from which t_s may be determined. Thermal resistance R is taken here from the inside surface to the outside t_o and

$$R = (0.01/0.16) + (0.110/0.19) + (0.025/0.035) + 0.18$$
$$+ (0.110/0.84) + 0.06$$
$$= 1.7268 \text{ m}^2\text{K/W}$$

Substituting:

$$5.13(18 - t_s) + 3(23 - t_s) = (t_s + 2)/1.7268$$

Thus

$$159 - 8.86t_s + 119 - 5.18t_s = t_s + 2$$

from which $t_s = 18.4$°C. The remaining interface temperatures can now be determined

$$(t_s - t_1)/R_p = (t_s - t_o)/R$$

Substituting:

$$(18.4 - t_1)/0.0625 = (18.4 + 2)/1.7268 = 11.814$$

from which $t_1 = 17.66$°C.

$$(t_s - t_2)/(R_p + R_c) = (t_s - t_o)/R$$

Substituting:

$$(18.4 - t_2)/(0.0625 + 0.579) = 11.814$$

from which $t_2 = 10.82°C$.

$$(t_s - t_3)/(R_p + R_c + R_i) = (t_s - t_o)/R$$

Substituting:

$$(18.4 - t_3)/(0.0625 + 0.579 + 0.714) = 11.814$$

from which $t_3 = 2.39°C$.

$$(t_s - t_4)/(R_p + R_c + R_i + R_a) = 11.814$$

Substituting:

$$(18.4 - t_4)/(0.0625 + 0.579 + 0.714 + 0.18) = 11.814$$

from which $t_4 = 0.26°C$.

$$(t_5 - t_o)/R_{so} = (t_s - t_o)/R$$

Substituting:

$$(t_5 + 2)/0.06 = 11.814$$

from which $t_5 = -1.29°C$.

SUMMARY FOR EXAMPLE 2.2

- The inside surface temperature of the external wall is obtained by separating the heat transfer coefficients for convection and radiation since t_r does not equate with t_{ai}.
  The heat flow path for the wall is shown in Figure 2.5.
- The dew point location in the wall should be checked. It should occur in the outer leaf where vapour can migrate to outdoors. A vapour barrier may be required on the hot side of the thermal insulation – that is, at the interface of the glass fibre slab and the inner leaf of the wall, to inhibit vapour flow from indoors to outdoors.

36 Heat conduction

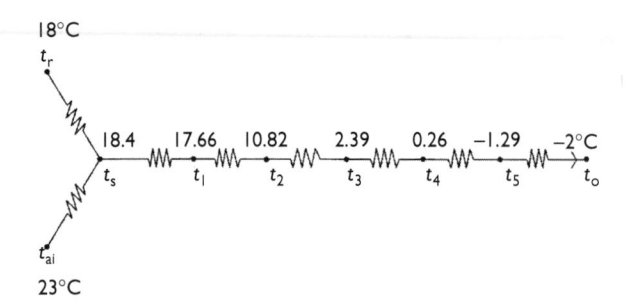

Figure 2.5 Example 2.2 – the heat flow path.

2.4 Heat conduction in ground floors

Heat loss through a solid floor in contact with the ground consists of two components: edge loss and ground loss. Edge loss is the more significant component and so rooms having ground floors with four exposed edges will have a greater heat loss than rooms with floors having fewer exposed edges.

The formula for the thermal transmittance coefficient U for solid floors in contact with the ground is given as:

$$U = (2k_e \cdot B)/(0.5b \cdot \pi) \text{artanh}(0.5b/(0.5b + 0.5w)) \text{ W/m}^2\text{K} \qquad (2.6)$$

where b = breadth (lesser dimension) of the floor in m,
w = thickness of surrounding wall taken to be 0.3 m,
k_e = thermal conductivity of earth = 1.4 W/mK. k_e depends upon the moisture content and ranges from 0.7 W/mK to 2.1 W/mK,
$B = \exp(0.5b/L_f) = (2.7183)^{(0.5b/L_f)}$,
L_f = length (greater dimension) of floor in metres.
artanh is one of the logarithmic forms of the inverse hyperbolic functions and is expressed in mathematical terms as:

$$\text{artanh}(x/a) = 0.5 \ln((a+x)/(a-x))$$

If another material is added to the composite structure, its thermal transmittance U_n can be adjusted thus:

$$U_n = 1/((1/U) + R_i) \text{ W/m}^2\text{K} \qquad (2.7)$$

where R_i is the thermal resistance of the added material.

Example 2.3 Calculating three thermal transmittance values through a ground floor

A floor in contact with the ground and having four exposed edges measures 20 m by 10 m.

(a) Determine the thermal transmittance for the floor.
(b) If the floor is surfaced with 15 mm of wood block having a thermal conductivity of 0.14 W/mK, determine its thermal transmittance coefficient.
(c) If the floor has a 25 mm thermal insulation membrane of expanded polystyrene (EPS) having a thermal conductivity of 0.035 W/mK in addition to its wood block finish, determine the transmittance coefficient.

Solution (a)
$B = \exp(0.5 \times 10/20) = \exp(0.25) = (2.7183)^{0.25} = 1.284$

From the mathematical expression of artanh:

$$\text{artanh}(5/(5+0.15)) = 0.5 \ln((5.15+5)/(5.15-5)) = 0.5 \ln(67.667)$$
$$= 2.1073$$

Substituting in equation 2.6

$$U = ((2 \times 1.4 \times 1.284)/(0.5 \times 10 \times \pi)) \times 2.1073 = 0.482 \text{ W/m}^2\text{K}$$

Solution (b)
From equation 2.7 which accounts for floor finish:

$$U_n = 1/((1/U) + R_i) = 1/((1/0.482) + (0.015/0.14))$$
$$= 1/(2.0747 + 0.107) = 0.46 \text{ W/m}^2\text{K}$$

Solution (c)
From equation 2.7 $U_n = 1/((1/0.46) + (0.025/0.035)) = 1/(2.174 + 0.714)$

$$= 0.346 \text{ W/m}^2\text{K}$$

SUMMARY FOR EXAMPLE 2.3

Floor data	U value (W/m²K)
Basic	0.482
Plus floor finish	0.46
Plus floor finish and insulation membrane	0.346

38 Heat conduction

Notes:

The floor structure does not play a part in the determination of the U value for floors in contact with the ground. The thermal conductivity of the earth k_e (equation 2.6) is accounted for.

The effect of reducing the number of exposed edges from four to two lowers the thermal transmittance by about half here.

The determination of the heat flux I to ground in practical heat loss calculations is based upon the indoor to outdoor design temperature differential and not indoor to ground temperature.

2.5 Heat conduction in suspended ground floors

A suspended ground floor above an enclosed air space is exposed to air on both sides. The air temperature below the floor will be higher than outdoor air temperature when it is at winter design condition because of the low rate of ventilation under the floor. The heat flow paths are shown in Figure 2.6(a) and (b). Figure 2.6(b) is the equivalent flow path which can assist in the determination of the thermal transmittance coefficient. The nomenclature for Figure 2.6 is as follows:

$$R_g = \text{thermal resistance through floor slab} = L/k \text{ m}^2\text{K/W}$$

R_{t1}, R_{t2} = transform resistances from delta to star arrangement

$$R_{t1} = R_r \cdot R_c/(R_r + 2R_c)$$

$$R_{t2} = (R_c)^2/(R_r + 2R_c)$$

$$R_r = 1/eh_r = 1/(0.9 \times 5.7) = 0.2 \text{ m}^2\text{K/W}$$

$$R_c = 1/h_c = 1/1.5 = 0.67 \text{ m}^2\text{K/W}$$

where the heat transfer coefficient h_c downwards at normal temperatures is 1.5 W/m²K. Thus

$$R_{t1} = (0.2 \times 0.67)/(0.2 + (2 \times 0.67)) = 0.09 \text{ m}^2\text{K/W}$$

$$R_{t2} = (0.67)2/(0.2 + (2 \times 0.67)) = 0.29 \text{ m}^2\text{K/W}$$

$$R_v = \text{ventilation resistance} = 0.63b$$

$$R_e = (1/U) - R_{si}$$

where the surface film resistance R_{si} downwards is taken as 0.14 m²K/W.

The thermal resistance R of the suspended ground floor following the heat flow path in Figure 2.6(b) can be shown as:

$$R = R_{si} + R_g + R_{t1} + [(1/(R_{t1} + R_e)) + (1/(R_{t2} + R_v))] - 1 \qquad (2.8)$$

Heat conduction 39

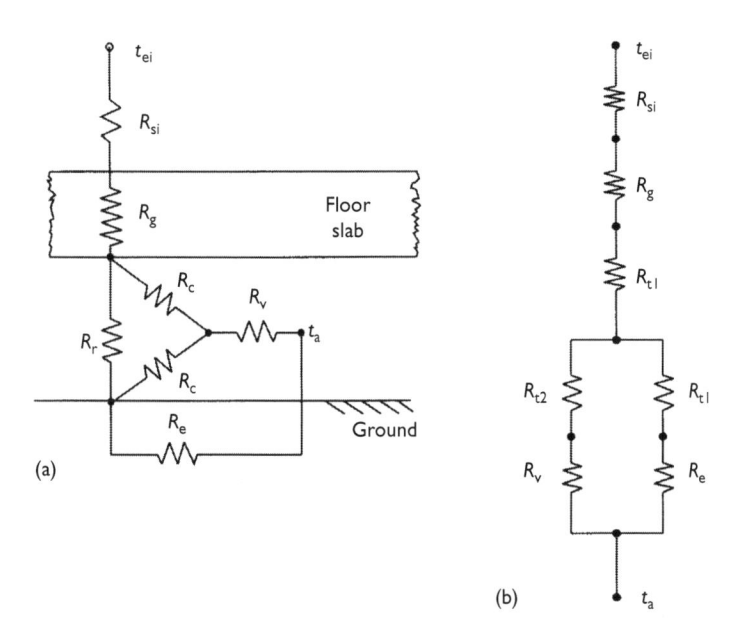

Figure 2.6 (a) Delta arrangement and (b) star arrangement, of thermal resistance network through suspended ground floors (Example 2.4).

Example 2.4 Determination of the thermal transmittance for a suspended ground floor

A suspended ground floor consists of 280 mm of cast concrete and measures 20 m by 10 m and has four exposed edges. Determine the thermal transmittance coefficient for the floor.

Solution
$R_g = L/k$ where the thermal conductivity of cast concrete from Table 2.1 is 1.4 W/mK. Thus

$R_g = 0.28/1.4 = 0.2$ m² K/W

$R_r = 0.2$ m²K/W from the text above

$R_c = 0.67$ m²K/W from the above text

$R_{t1} = 0.09$ m²K/W from above

$R_{t2} = 0.29$ m²K/W from above

The floor in this example is the same size as that in Example 2.3 and without a floor finish or thermal insulation membrane, the U value was

40 Heat conduction

calculated as 0.482 W/m²K. Make sure you have followed this procedure before continuing.

Taking the same conditions for the suspended ground floor in this example to find R_e:

$$R_e = (1/U) - R_{si} = (1/0.482) - 0.14 = 1.935 \text{ m}^2\text{K/W}$$

$$R_v = 0.63 \times 10 = 6.3 \text{ m}^2\text{K/W}$$

$$R_{si} = 0.14 \text{ m}^2\text{K/W from text above}$$

Substituting these values for the terms in equation 2.8

$$R = 0.14 + 0.2 + 0.09 + [(1/(0.09 + 1.935) + 1/(0.29 + 6.3)] - 1$$
$$= 0.43 + (0.494 + 0.152) - 1$$
$$= 0.43 + 1.55$$
$$= 1.98 \text{ m}^2\text{K/W}$$

Since $U = 1/R = 1/1.98 = 0.505$ W/m²K.

SUMMARY FOR EXAMPLE 2.4

- Note that the thermal transmittance coefficient for the floor in contact with the ground must be determined first. This U value is then used to determine R_e which is in equation 2.8 for suspended ground floors.
- If the floor has a 15 mm of wood block as a floor finish and a 25 mm thermal insulation membrane of EPS its thermal transmittance can be found adapting equation 2.7. Thus,

$$U_n = 1/((1/U) + R_w + R_i)$$

  and $U_n = 1/((1/0.505) + (0.015/0.14) + (0.025/0.035))$
  from which $U_n = 1/(1.98 + 0.107 + 0.714) = 0.357$ W/m²K.
- The comparison of the transmittance coefficients for the floor in contact with the ground (Example 2.3) and the suspended ground floor, each with four exposed edges can now be made:

Floor structure	Suspended ground floor	Floor in contact with ground
No insulation or finish	0.505	0.482
With insulation and finish	0.357	0.346

2.6 Thermal bridging and non-standard U values

External walls may not have a thermal transmittance which is consistent over the wall area. Structural columns may form thermal bridges in a cavity wall. At these points, the rate of conductive heat flow is high compared with that of the wall. The joisted flat roof is another example where the U value for the joisted part of the roof will be different to that for the spaces between the joists. That part of the structure having the higher U value is usually considered as the thermal bridge and will therefore cause the inside surface temperature to be at a lower value than the rest of the inner surface. Thermal bridges having high U values can cause discolouration of the inside surface and in extreme cases condensation.

There are three types of thermal bridges:

- Discrete bridges: These include lintels and structural columns which are flush with the wall or take-up part of the wall thickness.
- Multi-webbed bridges: These include hollow building blocks.
- Finned element bridges: Where the structural column protrudes beyond the width of the wall.

For walls with discrete bridges the average thermal transmittance

$$U = P_1 \cdot U_1 + P_2 \cdot U_2$$

where P_1 = unbridged area/total area and P_2 = bridged area/total area. Refer to Figure 2.7.

For finned element bridges, the bridged area in the calculation of P_2 includes the surface area of the protruding part of the thermal bridge. For a twin leaf wall with a discrete bridge in one of the leaves

$$U = 1/(R_b + R_h)$$

where

R_b = bridged resistance = $1/((P_1/R_1) + (P_2/R_2))$, and

$R_1 = R_{si} + L/k + 0.5R_a$

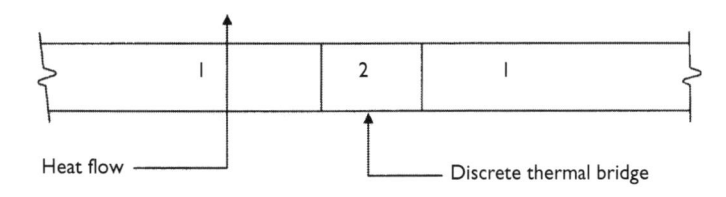

Figure 2.7 A thermally bridged wall.

$$R_2 = R_{si} + L/k + 0.5R_a \text{ (bridge)}$$

$$R_h = \text{homogeneous resistance} = 0.5R_a + L/k + R_{so}$$

Refer to Figure 2.8.

Example 2.5 Calculating the non-standard transmittance through a bridged external wall

Determine the non-standard U value for the bridged external wall shown in Figure 2.9.

Data: $R_{si} = 0.12$, $R_{so} = 0.06$, $R_a = 0.18$ m²K/W; $k_b = 0.84$, $k_1 = 0.19$, $k_2 = 1.4$ W/mK; $P_2 = 10\%$, $P_1 = 90\%$.

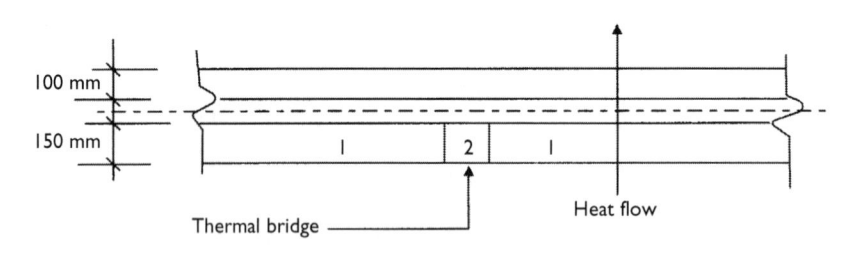

Figure 2.8 Cavity wall with bridge in inner leaf.

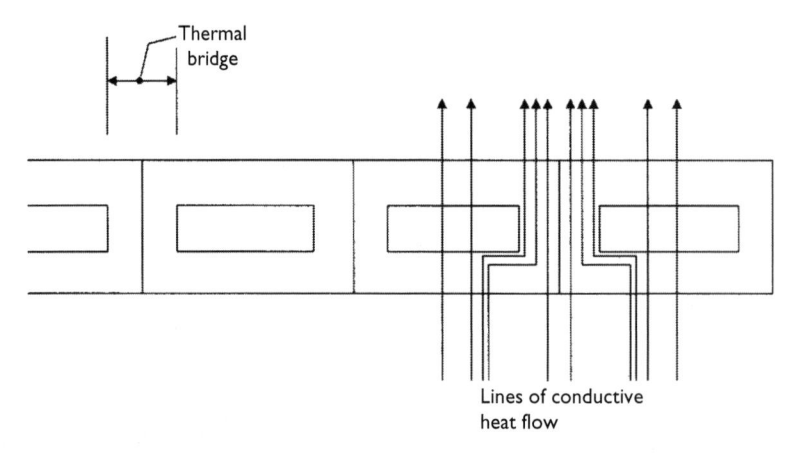

Figure 2.9 Non-uniform heat flow through a hollow block wall.

Solution
The data implies an external wall with concrete columns at intervals on the inner leaf. Adopting the formulae in the text above:

$$R_h = (0.5 \times 0.18) + (0.1/0.84) + 0.06 = 0.269 \text{ m}^2\text{K/W}$$

$$R_1 = 0.12 + (0.15/0.19) + (0.5 \times 0.18) = 0.9995 \text{ m}^2\text{K/W}$$

$$R_2 = 0.12 + (0.15/1.4) + (0.5 \times 0.18) = 0.317 \text{ m}^2\text{K/W}$$

$$R_b = 1/((0.9/0.9995) + (0.1/0.317)) = 0.822 \text{ m}^2\text{K/W}$$

Substituting for the non-standard U value:

$$U = 1/(0.822 + 0.269) = 0.917 \text{ W/m}^2\text{K}.$$

2.7 Non-standard *U* values, multi-webbed bridges

When considering the thermal transmittance for a hollow block the effect of lateral heat flow is significant. See Figure 2.9. An approximate calculation of the mean thermal resistance involves dividing the hollow block in two planes and employing the area weighted average thermal resistance (AWAR).

Consider the hollow block shown in Figure 2.10. By dividing the block into horizontal sections as shown in Figure 2.10, $R_a = 2L/k$.

$$\text{AWAR } A_1/R_b = (A_2/R_v) + (A_1 - A_2)/(L/k)$$

from which R_b is found and

$$R_c = R_a + R_b.$$

By dividing the block into vertical sections as shown in Figure 2.10, $R_a = L/k$ and $R_b = (2L/k) + R_v$

$$\text{AWAR } A_1/R_d = (A_1 - A_2)/R_a + A_2/R_b$$

from which R_d is found and $R_m = 0.5(R_c + R_d)$, and the non-standard thermal transmittance $U = 1/(R_{si} + R_m + R_{so})$. Clearly the hollow block will normally form part of the wall structure. If it is rendered on the outside and plastered on the inside, for example, the non-standard thermal transmittance will be

$$U = 1/(R_{si} + R_p + R_m + R_r + R_{so})$$

where R_p and R_r are the thermal resistances of the plaster and rendering, respectively.

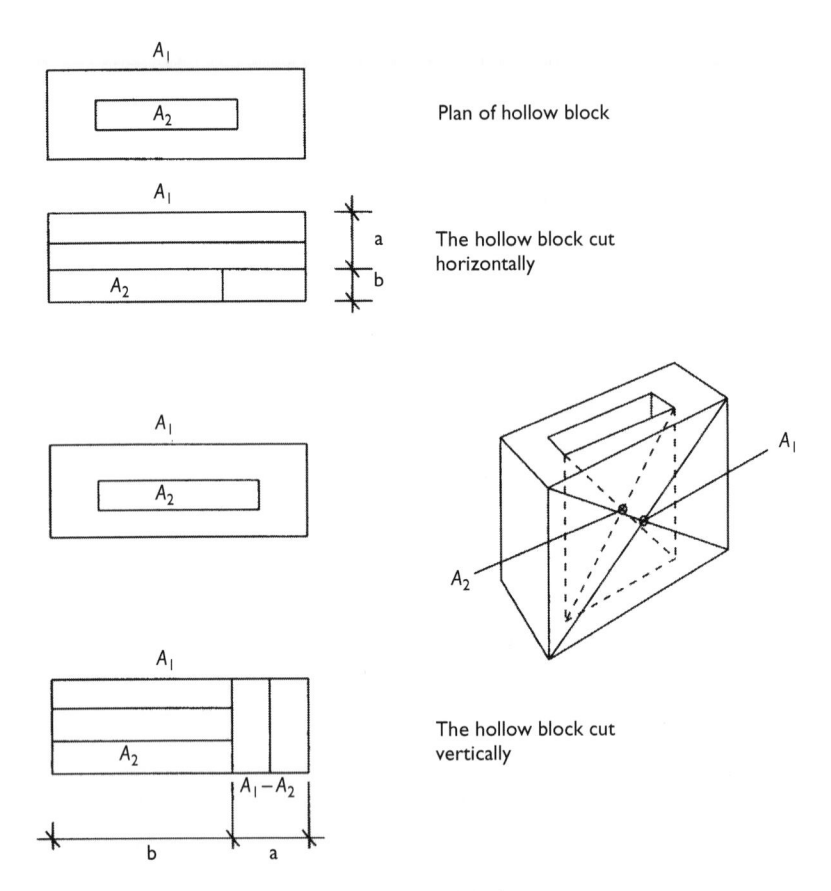

Figure 2.10 Thermal resistance of the hollow block.

Example 2.6 Determination of the non-standard transmittance through a hollow concrete block

Figure 2.11 shows an external hollow block wall rendered on the outer face and plastered on the inner face. From the data, determine the non-standard thermal transmittance coefficient for the wall and hence the heat flux, given indoor temperature is 20°C and outdoor temperature is −5°C.

Data: External wall specification – 15 mm of external rendering, medium weight hollow concrete block with air cavity filled with EPS, 16 mm of lightweight plaster. Thermal and film properties taken from Table 2.1.

Solution
You should now refer to Figure 2.10 which shows the way in which the block is cut for the purposes of determining the two thermal resistances

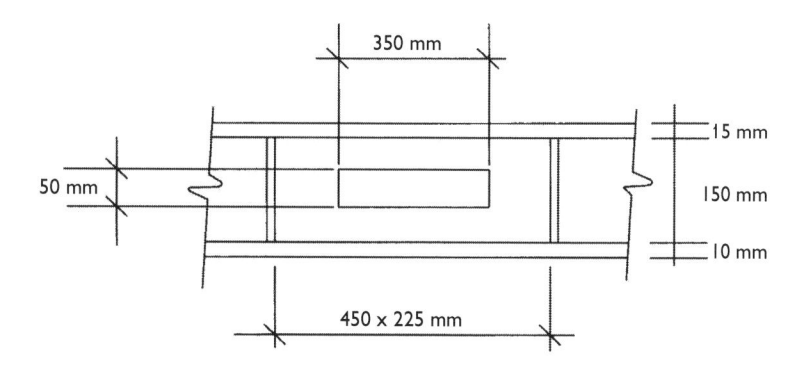

Figure 2.11 Hollow block wall (Example 2.6).

R_c and R_d. It is also important to identify the calculation procedure for the hollow block in the text above.

Slicing the block horizontally we have:

$$R_a = 2L/k = 2 \times 0.05/0.51 = 0.196$$

Hollow block dimensions and face and void surface areas are

$A_1 = 450 \times 225 = 101\,250$ this is approximately equivalent to 100

$A_2 = 350 \times 225 = 78\,750$ this is approximately equivalent to 79.

Thus

$$\text{AWAR } A_1/R_b = A_2/R_v + (A_1 - A_2)/(L/k).$$

Note that R_v is not a void/air cavity here as it is filled with EPS

$$100/R_b = (79/(0.05/0.035)) + ((100 - 79)/(0.05/0.51))$$

from which $R_b = 0.371$ m²K/W. Now $R_c = R_a + R_b = 0.196 + 0.371 = 0.567$ m²K/W.

Slicing the block vertically we have:

$$R_a = L/k = 0.15/0.51 = 0.294$$
$$R_b = (2L/k) + R_v = (0.1/0.51) + (0.05/0.035) = 1.6246.$$

Note again that R_v is not in this case a void/air cavity as it is filled with EPS

$$\text{AWAR } A_1/R_d = ((A_1 - A_2)/R_a) + A_2/R_b$$
$$100/R_d = ((100 - 79)/0.294) + (79/1.6246)$$

46 Heat conduction

from which $R_d = 0.833$ m²K/W. Now $R_m = 0.5(R_c + R_d) = 0.5(0.567 + 0.833) = 0.7$ m²K/W. For the composite wall

$$R = R_{si} + R_p + R_m + R_r + R_{so}$$
$$R = (0.12) + (0.01/0.16) + (0.7) + (0.015/0.5) + (0.06)$$
$$R = 0.9725 \text{ m}^2\text{K/W}.$$

The non-standard U value for the wall will therefore be:

$$U = 1/R = 1/0.9725 = 1.03 \text{ W/m}^2\text{K}.$$

The conductive heat flux $I = U \cdot dt = 1.03 \times (20 + 5) = 25.75$ W/m² of wall.

SUMMARY FOR EXAMPLE 2.6
This is an average rate of heat flow through the wall. Thermal bridges are formed between the cavities filled with EPS and may result in discolouration on the inside surface of the plaster due to surface temperature variations along the wall. A thermal bridge is a path located through a structure where the rate of heat flow is substantially increased as a result of the materials used.

2.8 Radial conductive heat flow

For plane (flat) surfaces, surface area A is constant and $Q = U \cdot A \cdot dt$ Watts. For cylinders and spheres the surface area is not constant either for multiple layers of material or for single layers having a measurable thickness. Thus as the radius increases through the thermal insulation material surrounding a pipe transporting hot or chilled water, for example, so does the surface area of the insulation surrounding that pipe.

If $A_m =$ the mean surface area of each layer of thermal insulation around the pipe, then from Fourier's equation for a single layer:

$$Q = k \cdot A_m \cdot dt/L \text{ Watts}$$

This can be rewritten as $Q = dt/(L/k \cdot A_m)$. Thus for multiple layers of insulation around a pipe:

$$Q = dt/\sum(L/k \cdot A_m) \text{ Watts}$$

where $dt =$ temperature differential between the inside pipe surface and the outside insulation surface.

If dt is taken from the fluid flowing inside the pipe to the outside air:

$$Q = dt/((1/A_i \cdot h_{si}) + \sum(L/k \cdot A_m) + (1/A_o \cdot h_{so})) \qquad (2.9)$$

A_m is the logarithmic mean area of each layer of material and A_i and A_o are the inside and outside surface areas, respectively.

For cylinders

$$A_m = (\pi(d_2 - d_1)L)/\ln(d_2/d_1) \tag{2.10}$$

You can see here the similarity with the surface area of a cylinder $A = \pi \cdot L \cdot d$. Furthermore, A_m is approximately equal to $\pi((d_1 + d_2)/2)L$. You should confirm this in Example 2.7 below.

For spheres,

$$A_m = \pi \cdot d_1 \cdot d_2 \text{ for each thermal insulation layer} \tag{2.11}$$

This is one of the two methodologies for introducing radial heat flow in pipes and circular ducts having one or more layers of thermal insulation.

Example 2.7 Calculating the rate of heat loss from a cylindrical thermal storage vessel

A cylindrical vessel 4 m diameter and 7 m long has hemispherical ends giving it an overall length of 11 m. The vessel which stores water for space heating at 85°C is covered with 300 mm of lagging which has a thermal conductivity of 0.05 W/mK. Determine the heat loss from the vessel to the plant room which is held at 22°C. Take the outside surface heat transfer coefficient as 12 W/m²K and ignore the influence of the vessel wall and the inside heat transfer coefficient.

Solution
Since the inside heat transfer coefficient and the vessel wall thickness is not accounted for, equation 2.9 for cylinders needs adapting and:

$$Q = dt/((L/(kA_m) + (1/h_{so}A_o))$$

where $dt = (85 - 22)$ and from equation 2.10,

$$A_m = \pi(4.6 - 4.0)7/\ln(d_2/d_1) = 94.38 \text{ m}^2$$
$$A_o = \pi \times 4.6 \times 7 = 101.16 \text{ m}^2$$

Substituting we have:

$$Q = (85 - 22)/(0.3/0.05 \times 94.38) + (1/(12 \times 101.16))$$
$$Q = 63/(0.0636 + 0.0008)$$

You can see that the effect of the last term involving h_{so} and A_o is insignificant. Thus for the cylinder $Q = 978$ W. The hemispherical ends of the vessel

48 Heat conduction

form a sphere, for the logarithmic mean area A_m of which, equation 2.11 can be used and:

$$A_m = \pi \times 4.0 \times 4.6 = 57.8 \text{ m}^2$$

Ignoring the effect of h_{so}, the heat loss from the hemispherical ends of the vessel will be:

$$Q = dt/(L/k \cdot A_m)$$

Substituting:

$$Q = (85 - 22)/(0.3/0.05 \times 57.8) = 607 \text{ W}$$

The total heat loss from the vessel to the plant room $= 978 + 607 = 1585$ W.

The classical methodology for developing a formula for radial heat flow integrates Fourier's law for one-dimensional heat flow.
Thus

$$Q = -k \cdot A \cdot dt/dr \text{ W for a single layer of material}$$

Considering unit length of a cylinder L where $A = 2\pi r L$,

$$Q/L = -k(2\pi r) \, dt/dr \text{ W/m run}$$

If temperatures are steady Q remains steady and the temperature gradient from the fluid flowing in the pipe to surrounding air decreases with increasing radius r. Integrating between the limits r_1 and r_2, refer to Figure 2.12:

$$\int (Q/L) dr/r = -\int \pi k \, dt$$

$$(Q/L) \cdot \ln(r_2/r_1) = -2\pi k(t_1 - t_2)$$

from which $Q/L = -2\pi k(t_1 - t_2)/\ln(r_2/r_1)$ W/m for a single layer of material. The minus sign indicates a heat loss and may be ignored.

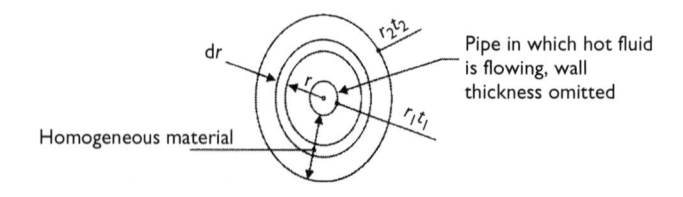

Figure 2.12 Radial heat flow integrating between the limits of r_1 and r_2.

For a multilayer cylindrical wall:

$$Q/L = (2\pi \, dt)/[(1/r_1 \cdot h_{si}) + ((\ln r_2/r_1)/k_2) + ((\ln r_3/r_2)/k_3) \\ + \cdots + (1/r_n \cdot h_{so})] \text{ W/m run} \quad (2.12)$$

Example 2.8 Determining the rate of heat loss from an insulated steam pipe

A 50 m length of steam pipe connects two buildings and carries saturated steam at 29 bar gauge. The pipe has an internal diameter of 180 mm, a wall thickness of 19 mm and is covered by two layers of thermal insulation. The inner layer is 20 mm thick and the outer layer is 25 mm thick. The thermal effect of the outer protective casing to the pipe insulation can be ignored. Determine the rate of heat loss from the pipe.

Data: Outdoor temperature = 5°C. Thermal conductivity of pipe wall $k_w = 48$, inner layer of insulation $k_i = 0.035$, outer layer of insulation $k_o = 0.06$ W/mK. Heat transfer coefficient at the pipe inner surface $h_{si} = 550$, heat transfer coefficient at the outer layer insulation surface $h_{so} = 18$ W/m²K.

Solution
Refer to Figure 2.13, $r_1 = 90$, $r_2 = 109$, $r_3 = 129$ and $r_4 = 154$ mm. From the tables of *Thermodynamic and Transport Properties of Fluids* the temperature of saturated steam at 30 bar abs. = 234°C. Substituting data into equation 2.12,

$$Q/L = 2\pi(234 - 5)/\{[(1/(0.09 \times 550)) + (\ln 109/90) + (\ln 129/109) \\ + (\ln 154/129) + (1/0.154 \times 18)]\}$$

Note the ratios of r_2/r_1 etc. are kept in mm for convenience without loss of integrity.

$$Q/L = 1439/(0.002 + 0.004 + 4.813 + 2.952 + 0.361)$$
$$= 1439/8.132 = 177 \text{ W/m run}$$
$$Q = 177 \times 50 = 8847 \text{ W}$$

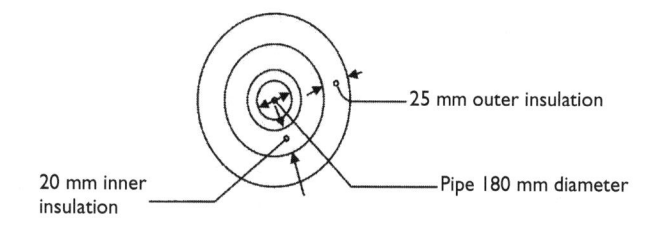

Figure 2.13 Insulated steam pipe (Example 2.8).

50 Heat conduction

SUMMARY FOR EXAMPLE 2.8
You can see that the effect of the inside heat transfer coefficient h_{si} and the pipe wall is insignificant and therefore frequently ignored in the solution to this type of problem.

Example 2.9 Calculating the thickness of thermal insulation for a pipe

Determine from the data the thickness of thermal insulation to be applied to a pipe conveying water over a distance of 100 m if its temperature is not to fall below 79°C.

Data:

Outdoor temperature	−1°C
Initial water temperature	80°C
Water flow rate	1.5 kg/s
Outside diameter of the pipe	66 mm
Specific heat capacity of water	4200 J/kg/K
Thermal conductivity of the insulation	0.07 W/mK
Coefficient of heat transfer at the outside surface of the insulation	10 W/m²K

Solution
The effect of the heat transfer coefficient at the inside surface of the pipe can be ignored. The effect of the thickness of the pipe wall may be ignored. The temperature of the outer surface of the pipe can therefore be taken as 80°C at the beginning of the pipe run and 79°C at the end of the run. Now the maximum heat loss from the pipe will be:

$$Q/L = M \cdot C \cdot dt/L = (1.5 \times 4200 \times (80 - 79))/100 = 63 \text{ W/m run}$$

The mean temperature of the pipe surface $= (80 + 79)/2 = 79.5°C$.
Let $z =$ insulation thickness in metres. Adopting equation 2.12,

$$63 = (2\pi(79.5 + 1))/[\ln((0.033 + z)/(0.033)/0.07)$$
$$+ (1/10(0.033 + z)]$$

Collecting the common factors on the left-hand side and inverting the formula:

$$161\pi/63 = [\ln((0.033 + z)/0.07)] + [1/10(0.033 + z)]$$

This may be written as

$$8.03 = [Y] + [W]$$

Table 2.2 Results for solution to Example 2.9

z	0.033 + z	Y	W	(Y + W)
0.01	0.043	3.75	2.33	6.08
0.012	0.045	4.44	2.22	6.66
0.014	0.047	5.07	2.13	7.20
0.016	0.049	5.65	2.04	7.69
0.018	0.051	6.22	1.96	8.18
0.020	0.053	6.78	1.89	8.67

If values are now given to z then the terms Y and W in the formula can be reduced to numbers. The results are given in Table 2.2.

From the tabulated results in Table 2.2, it can be observed that the insulation thickness lies between 16 and 18 mm for the equation, $8.03 = (Y + W)$ to balance. If z and $(Y + W)$ are plotted on a graph the thickness may be obtained as 17.3 mm. In practice, 20 mm thick pipe insulation would no doubt be selected.

SUMMARY FOR EXAMPLE 2.9

- The maximum heat loss from the pipe is 63 W/m run. This may be converted to a permitted heat loss expressed in W/m^2:

$$I = 63/(\pi \cdot d \cdot L) = 63/(\pi \times 0.066 \times 1.0) = 304 \text{ W/m}^2$$

- It is suggested that you now undertake a similar calculation to find the minimum thickness of thermal insulation for the pipe given a maximum permitted heat loss of 200 W/m^2, when outdoor temperature is $-1°$C.

The following start may help:

$$200 \text{ W/m}^2 = (200 \times \pi \times 0.066 \times 1.0) = 41.47 \text{ W/m run of pipe}$$

Since $Q/L = M \cdot C \cdot dt/L$,

$$dt = ((Q/L) \times L)/M \cdot C$$

thus

$$dt = (41.47 \times 100)/(1.5 \times 4200) = 0.66 \text{ K}$$

This means that the temperature of the pipe/water at the end of the run will be $80 - 0.66 = 79.34°$C and the mean temperature will be $(80 + 79.34)/2 = 79.67°$C. This data can now be substituted into the formula and values given to z, the insulation thickness. The solution comes to 36 mm of pipe insulation. You should now confirm this is so.

52　Heat conduction

Example 2.10　Determining the minimum thickness of thermal insulation required to prevent surface condensation

A sheet steel duct 600 mm in diameter carries air at $-25°C$ through a room held at $20°C \cdot db$ and 68% saturated to a food freezing processor. Determine the minimum thickness of thermal insulation required to prevent condensation occurrence on the outer surface of the duct insulation. Recommend a suitable finish to the duct insulation.

Data: Atmospheric pressure 101 325 Pa; heat transfer coefficient at the outer surface 8 W/m²K; thermal conductivity of duct insulation $k = 0.055$ W/mK. Ignore the effect of the heat transfer coefficient at the inner surface of the duct and the duct thickness.

Solution

From hygrometric data, dew point temperature t_d of air at $20°C \cdot db$ and 68% saturated is $14°C$. To avoid the incidence of condensation the outer surface of the thermal insulation must not fall below this temperature.

A heat balance may be drawn up, assuming steady temperatures, and the heat flow from the room to the outer surface of the duct insulation = the heat flow from the insulation outer surface to the air in the duct.

Let z = minimum thickness of insulation in metres and adopting the classical formula 2.12 for radial heat flow and generic equation 2.3, modified for the heat balance. Thus

$$2\pi(t_i - t_{so})/(1/h_{so}(0.3 + z)) = 2\pi(t_{so} - t_a)/[\ln((0.3 + z)/0.3)/k]$$

Simplify and substitute:

$$(20 - 14)/(1/8(0.3 + z)) = (14 + 25)/[\ln((0.3 + z)/0.3)/0.055]$$

Rearranging the equation:

$$6 = \{(39/[\ln((0.3 + z)/0.3)/0.055](1/8(0.3 + z))\}$$

This is a simple relationship of $6 = [Y] \times [W]$.
Giving values to insulation thickness z and tabulating in Table 2.3.

Table 2.3　Results for solution to Example 2.10

z	(0.3 + z)	W	Y	(YW)
0.02	0.32	0.391	33.23	12.99
0.03	0.33	0.379	22.51	8.53
0.04	0.34	0.368	17.40	6.31
0.045	0.345	0.362	15.35	5.55

SUMMARY FOR EXAMPLE 2.10
From Table 2.3 it is seen that the minimum thickness of duct insulation is about 42 mm. The surface finish to the duct insulation must be waterproof to ensure against the migration of water or vapour into the thermal insulation from the surrounding air.

2.9 Chapter closure

You have now been introduced to the principles of conductive heat flow and applied these principles to practical applications relating to

> heat flow through external walls;
> heat flow through ground floors;
> heat flow resulting from non standard U values;
> radial heat flow from thermally insulated pipes, ducts and vessels.

You are familiar with the potential errors in the determination of building heat loss when inequalities between indoor air temperature and mean radiant temperature are ignored and the fact that errors are most likely where buildings are not thermally insulated or sealed to current building regulation standards.

Chapter 3

Heat convection

Nomenclature

C	specific heat capacity (kJ/kgK)
dt	temperature difference (K)
dt_m	log mean temperature difference (K)
Gr	Grashof number
h_c	heat transfer coefficient for convection (W/m^{2}K)
h_{si}/h_{so}	heat transfer coefficients at the inside/outside surface (W/m^{2}K)
I	heat/cooling flux (W/m^2)
k	thermal conductivity (kW/mK or W/mK)
M	mass flow rate (kg/s)
Nu	Nusselt number
Pr	Prandtl number
Q	heat/cooling rate (W)
Re	Reynolds number
R_f	fouling resistance (m^{2}K/W)
$T, T_m, T_s, T_a, T_1, T_2$	absolute temperature (K)
t_s, t_f	customary temperature (°C)
u	mean velocity (m/s)
U	thermal transmittance coefficient (W/m^{2}K)
v	specific volume (m^3/kg)
x, d, D	characteristic dimensions (m)
β	coefficient of cubical expansion (K^{-1})
μ	dynamic viscosity (kg/m·s)
ν	kinematic viscosity (m^2/s), specific volume (m^3/kg)
ρ	density (kg/m^3)

3.1 Introduction

Heat convection can occur in liquids and gases when the molecules move freely and independently. This occurs as a result of cubical expansion or contraction of the fluid when it is heated or cooled causing changes in fluid density. This initiates movement by natural means with the warmer fluid rising and the cooler fluid dropping, due to the effects of gravity upon it. This type of movement is termed free convection and occurs, for example, over radiators and natural draught convectors. See Figure 3.1.

Forced convection is obtained with the aid of a prime mover, such as a pump or fan, and occurs at the heat exchanger of a fan coil unit, for example, with pumped water flowing inside the heat exchanger pipes and fan assisted air flowing over the finned heat exchanger surface. See Figures 3.2 and 3.3.

Free convection relies on natural forces and its effectiveness in heat transfer relies on

- velocity of the secondary fluid over the heat exchanger surface;
- magnitude of the temperature difference between the primary and secondary fluids;
- size and shape of heat exchanger and its position in space. Refer to equations 3.5–3.9.

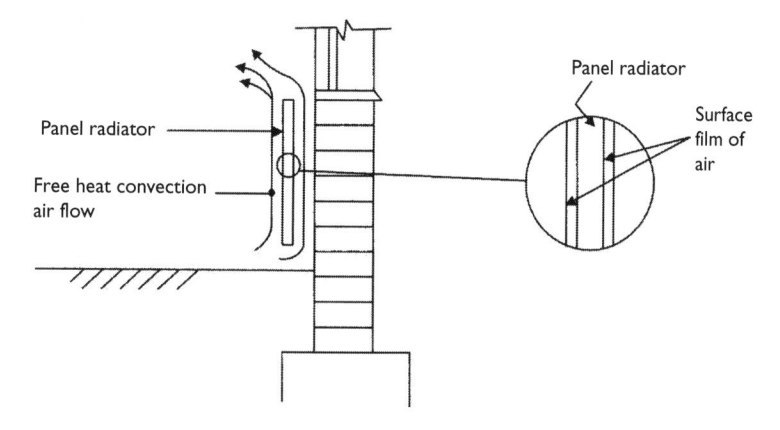

Figure 3.1 Air flow in free convection over panel radiator.

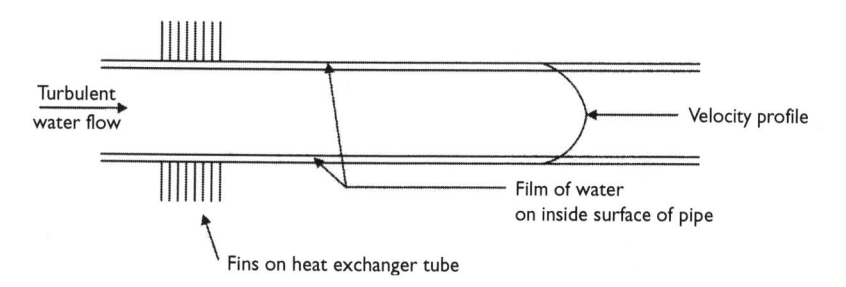

Figure 3.2 Water flow inside finned tube heat exchanger providing forced convection.

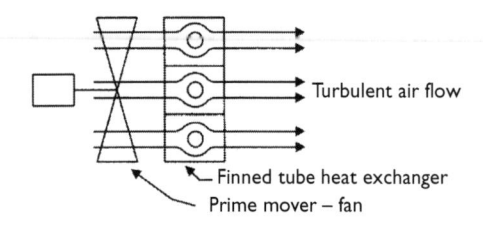

Figure 3.3 Air flow in forced convection over a finned tube heat exchanger.

The velocity of the secondary fluid over the heat exchanger surface will influence

- type of flow whether laminar or turbulent;
- the leaving temperature of the secondary fluid off the heat exchanger;
- the thickness of the film at the interface of the moving secondary fluid and the heat exchanger surface on the convective heat transfer;
- the degree of contact between the secondary fluid and the heat exchanger surface. This is accounted for by the contact factor which is 1.0 for complete contact around the secondary surface by the secondary fluid. Usually, the contact factor is less than 1.0 and is likely to be at its maximum in turbulent flow.

Forced convection, on the other hand, is less affected by the shape of the heat exchanger or its position in space. It has a positive and directional movement and is not so subject to natural forces. However, most applications of forced convection rely on it as the main mode, if not the only mode, of heat transfer. In the case of unit heaters or fan coil units in heating mode, which are mounted at high level/ceiling level, it is important to account for the fact that leaving air temperatures make the air buoyant and this buoyancy must be overcome to ensure that it reaches the working plane.

There are a number of properties of a flowing fluid which influence heat convection: β, μ, ν, ρ, k, C, T. There are other quantities that are affected or which affect heat convection: M, dt, h_c, u. These variables together with appropriate characteristic dimensions may be collected in dimensionless groups by analysis as shown here. Refer also to Appendix 2 (Examples A2.5, A2.6)

$$\text{Reynolds number,} \quad Re = \rho u d/\mu = dM/\mu A \tag{3.1}$$

$$\text{Nusselt number,} \quad Nu = h_c \cdot d/k \tag{3.2}$$

$$\text{Prandtl number,} \quad Pr = \mu C/k \tag{3.3}$$

$$\text{Grashof number,} \quad Gr = [\beta(\rho^2)(x^3)dt \cdot g]/\mu^2 \tag{3.4}$$

Notes:

1 Cubical expansion of gases, $\beta = 1/T_m$ K^{-1}, where $T_m = (T_s + T_a)/2$.
2 Properties of air and water at different temperatures can be obtained from the tables of *Thermodynamic and Transport Properties of Fluids* (SI units) and include: C, k, μ, ρ, v, Pr.
3 Density $\rho = 1/v$.
4 Before adopting a formula for the purposes of solving a problem it is necessary to determine whether the fluid is in laminar flow or turbulent flow. Refer to Chapter 6 for a detailed analysis.

3.2 Rational formulae for free and forced heat convection

The formulae given in this section have been determined by dimensional analysis (see Appendix 2) and the associated constants and indices have been determined empirically which is to say by practical experiment. The solutions resulting from application of these formulae must therefore be treated as approximate but sufficiently accurate. Because of the complexity of determining the heat transfer coefficient for convection h_c from a surface, specific formulae have been developed for different shapes of surface in various positions in space. Some of the more appropriate of these are given here. The following Grashof and Reynolds numbers identify the type of fluid flow and associated formulae are given.

Free convection of air over vertical plates:

Laminar flow, $Gr < 10^8$

Turbulent flow, $Gr > 10^9$

Laminar flow, $(Nu)_x = 0.36[(Gr)_x]^{0.25}$ (3.5)

Turbulent flow, $(Nu)_x = 0.13[(Pr)(Gr)_x]^{0.33}$ (3.6)

Free convection of air over horizontal plates:

Laminar flow, $1.4 \times 10^5 < (Gr) < 3 \times 10^7$

Turbulent flow, $3 \times 10^7 < (Gr) < 3 \times 10^{10}$

For hot surfaces looking up and cool surfaces looking down:

Laminar flow, $h_c = 1.4[(t_s - t_f)/D]^{0.25}$ (3.7)

For hot surfaces looking down and cool surfaces looking up:

$$\text{Laminar flow,} \quad h_c = 0.64[(t_s - t_f)/D]^{0.25} \tag{3.8}$$

where $D =$ (length + width)$/2$.
For hot surfaces looking up and cool surfaces looking down:

$$\text{Turbulent flow,} \quad h_c = 1.7(t_s - t_f)^{0.33} \tag{3.9}$$

Forced convection:
Turbulent flow inside tubes where $Re > 2500$,

$$(Nu)_d = 0.023[(Re)_d]^{0.8}(Pr)^{0.33} \tag{3.10}$$

Turbulent flow outside tube bundles where $Re > 4 \times 10^5$,

$$(Nu)_d = 0.44[(Re)_d]^{0.55}(Pr)^{0.31} \tag{3.11}$$

Turbulent flow over flat plates where $Re > 1 \times 10^5$,

$$(Nu)_x = 0.037[(Re)_x]^{0.8}(Pr)^{0.33} \tag{3.12}$$

Free convection over horizontal cylinders:
Laminar flow where $Gr < 10^8$ for air and water,

$$(Nu)_d = 0.53[(Gr)_d(Pr)]^{0.25} \tag{3.13}$$

Free convection over vertical cylinders:
Turbulent flow where $Gr > 10^9$

$$\text{for air,} \quad Nu = 0.1[(Gr)(Pr)]^{0.33} \tag{3.14}$$

$$\text{for water,} \quad Nu = 0.17[(Gr)(Pr)]^{0.33} \tag{3.15}$$

Analysis of equation 3.6 is included in Appendix 2, Example A2.6. Analysis of equation 3.10 is included in Appendix 2, Example A2.5.
From the above text you will have seen that:

- Free convection is related to the Grashof number Gr, where laminar and turbulent flow depends upon its magnitude.
- Forced convection is related to the Reynolds number Re, where laminar and turbulent flow depends upon its magnitude.

From the Nusselt formula the heat transfer coefficient for convection h_c is determined and convective heat transfer is obtained from:

$$Q = h_c \times A \times dt \text{ Watts} \tag{3.16}$$

where dt = surface temperature t_s minus fluid temperature t_f and A = area of heat exchanger surface.

The heat transfer coefficient for convection h_c is dependent upon the magnitude of the difference between the heat exchanger surface temperature and the bulk fluid temperature and also upon the thickness of the laminar sublayer on the heat exchanger surface. Refer to Chapter 6.

3.3 Temperature definitions

Various terms are used for temperature and temperature difference in convective heat transfer. They include

- mean bulk temperature which refers to the arithmetic mean temperature of the fluid flowing;
- mean film temperature which refers to the mean temperature of the bulk plus heat exchanger temperature;
- mean temperature difference which refers to the difference between heat exchanger and mean bulk temperature;
- log mean temperature difference (LMTD) – if the temperature of both fluids vary, true temperature difference will be the logarithmic mean value.

Example 3.1 Calculating mean bulk and mean film temperature, and mean temperature difference for a given rise in temperature of the secondary fluid

For a fluid flowing, given a rise in temperature of 70–80°C as a result of a constant heat exchange temperature of 140°C. Find the mean bulk temperature, the mean film temperature of the fluid and the mean temperature difference.

Solution
A heat exchanger whose surface is at a constant temperature is usually associated with the use of steam or refrigerant which gives up its latent heat in the heat exchanger at a constant temperature, and in the process changes its state.

Mean bulk temperature of the fluid flowing $= (70 + 80)/2 = 75°C$.
Mean film temperature of the fluid flowing $= 0.5((70+80)/2)+140) = 107.5°C$.
Mean temperature difference between the fluid flowing and the heat exchanger surface $= (140 - (70 + 80)/2) = 65$ K.

60 Heat convection

Example 3.2 Calculating mean bulk temperature and mean film temperature for primary and secondary fluids

A heat exchanger carries high temperature hot water at inlet and outlet temperatures of 150 and 110°C. The secondary fluid rises in temperature from 70 to 82°C. Determine the mean bulk temperatures and mean film temperature. Comment on the mean temperature difference between the fluids.

Solution

Mean bulk temperature of the secondary fluid = $(70 + 82)/2 = 76°C$.
Mean bulk temperature of the primary fluid = $(150 + 110)/2 = 130°C$.
Mean film temperature of both fluids = $0.5(76 + 130) = 103°C$.

As both the fluid temperatures vary, knowledge of the *mode* of flow of the primary and secondary fluids is required together with the application of the LMTD. See Example 3.3 below.

Example 3.3 Finding the LMTD in counterflow

A counterflow heat exchanger carries high temperature hot water at inlet and outlet temperatures of 150°C and 110°C. The secondary fluid rises in temperature from 70°C to 82°C. Find the LMTD between the primary and secondary fluids.

Solution
This topic is considered in detail in Chapter 9. LMTD, dt_m, accounts for temperature variations in both the primary and secondary fluids and from Chapter 9:

$$dt_m = (dt_{max} - dt_{min})/\ln(dt_{max}/dt_{min})$$

For counterflow:

Primary   150   →   110
Secondary   82   ←   70
           68 dt_{max} 40 dt_{min}

Substitute:

$$dt_m = (68 - 40)/\ln(68/40) = 52.77 \text{ K}.$$

SUMMARY FOR EXAMPLE 3.3
It is interesting to note that the arithmetic mean = $[((150 + 110)/2) - ((82 + 70)/2)] = (130 - 76) = 54$ K, which is not much different to the logarithmic mean. If the high temperature return is at 120°C instead of

110°C, the LMTD is calculated as 58.5 K whereas the arithmetic mean is 59 K. Here the two values are even closer. You should now confirm that this is so.

If conductive heat flow is considered from the hot fluid across the exchanger wall to the cold fluid, the overall U value for the heat exchanger is appropriate:

$$U = 1/((1/h_{si}) + R_f + (1/h_{so})) \; W/m^2 K \tag{3.17}$$

Hence, $Q = U \cdot A \cdot dt$ W (3.18)

where dt = temperature of hot fluid minus the temperature of the cold fluid.

Notes:

- The thermal resistance of the heat exchanger wall is ignored since it is insignificant.
- The overall U value is dependent upon the thickness of the laminar sublayer on both the inside and the outside of the heat exchanger and hence on the type of fluid flow. See Chapter 6.
- R_f is the fouling factor measured in $m^2 K/W$. If maintenance of the heat exchanger is done on a regular basis it is sometimes ignored. It can range from 0.09 to 0.2 $m^2 K/W$. Refer also to Section 9.3 and Example 9.4.

There now follows some practical examples which adopt the formulae introduced in the text above. You should follow them by noting the solution procedure in each case.

3.4 Convective heat output from a panel radiator

Example 3.4 Determining the convective heat transfer from a panel radiator

Low temperature hot water flows through a single panel vertically mounted radiator 1 m high by 1.5 m long. The mean temperature of the circulating water is 60°C and the temperature of the surrounding air is 19°C.

Determine the heat transferred by convection. Ignore the resistance of the air and water films on each side of the radiator and the radiator material. Evaluate the properties at the mean film temperature.

Solution
The first step in the solution procedure is to determine the Grashof number which will establish whether the air flow over the radiator is in the laminar or turbulent region. The mean film temperature = $0.5(60 + 19) = 39.5°C = 312.5$ K.

62 Heat convection

Referring to the Grashof number (equation 3.4), cubical expansion, $\beta = 1/(312.5) = 0.0032$. Interpolating from the tables of *Thermodynamic and Transport Properties of Fluids*, air density at mean absolute temperature of 313 K, $\rho = 1.13\,\text{kg/m}^3$. Air viscosity $\mu = 0.000019$ kg/m s, vertical height $x = 1.0$ m and temperature difference between the radiator and air $dt = 60 - 19 = 41$ K.

Now substituting into the Grashof number:

$$Gr = [0.0032 \times (1.13)^2 \times (1.0)^3 \times 41 \times 9.81]/(0.000019)^2$$

$$Gr = 4.55 \times 10^9$$

Note that x is the height of the panel; a lower height yields a lower Gr and ultimately a lower heat transfer coefficient h_c. A long low radiator will therefore give a lower convective output than a short tall radiator of the same area. This is caused by the increased stack effect that the taller radiator has which induces greater vertical air flow over its surface.

From the text, turbulent flow in free convection over vertical plates commences when $Gr > 10^9$ and therefore applies here. The adopted formula will be equation 3.6. The Prandtl number can be calculated or the value interpolated from the tables of *Thermodynamic and Transport Properties of Fluids* at the mean air temperature of 313 K in which case, $Pr = 0.703$.

Note that the tables quote the thermal conductivity k in kW/mK and specific heat capacity C in kJ/kgK. By interpolating μ, C and k for dry air from the tables, Pr can also be calculated from equation 3.3. From equation 3.6

$$(Nu)_x = 0.13((Pr)(Gr)_x)^{0.33}$$

Substituting:

$$(Nu)_x = 0.13((0.703)(4.55 \times 10^9))^{0.33}$$
$$= 178$$

But from equation 3.2 $Nu = h_c \cdot x/k$ where x is a characteristic dimension. Here x is the panel height thus $x = 1.0$ and

$$h_c = Nu \cdot k/x = 178 \times 0.0273/1.0 = 4.86\,\text{W/m}^2\text{K}$$

The surface area of the radiator is $1.0 \times 1.5 \times 2 = 3.0$ m^2. From equation 3.16, free heat convection

$$Q = h_c \cdot A \cdot dt = 4.86 \times 3 \times (60 - 19) = 598\,\text{W}.$$

SUMMARY FOR EXAMPLE 3.4

As already mentioned in the solution, the height x of a radiator influences its heat transfer by free convection by affecting the Grashof number. The lower the radiator height the lower is the convective heat transfer for the same surface area. This is confirmed from reference to manufacturers' literature. In practical terms, the 'stack effect' of the free convection over the radiator surface increases with its height – thus increasing convective output.

3.5 Heat output from a freely suspended pipe coil

Example 3.5 Calculating the heat output from a greenhouse pipe coil

A freely suspended 100 mm bore horizontal pipe is located at low level in a greenhouse to provide heating. It has a surface emissivity of 0.9 and is supplied with water at 85°C flow and 73°C return. The greenhouse is held at an air temperature of 19°C and a mean radiant temperature of 14°C. Evaluating the properties at the mean film temperature determine the heat output from the pipe given that it is 15 m in length. Take the outside diameter as 112 mm.

Solution

The Grashof number will determine the type of air flow over the pipe.

The mean film temperature $= 0.5\{[(85 + 73)/2 + 19]\} = 49°C = 322\,K$

Using the data for dry air from the tables for *Thermodynamic and Transport Properties of Fluids*,

$C = 1.0063$ kJ/kgK, $\mu = 0.00001962$ kg/ms, $\rho = 1.086$ kg/m³, $k = 0.00002816$ kW/mK, $dt = 79 - 19 = 60$ K, $x = d = 0.112$ m. Also $\beta = 1/T_m = 1/322 = 0.0031$

Substituting into equation 3.4

$Gr = [0.0031 \times (1.086)^2 \times (0.112)^3 \times 60 \times 9.81]/(0.00001962)^2$

$Gr = 7.87 \times 10^6$

This identifies laminar air flow over the pipe and from the tables at the mean film temperature of 322 K, $Pr = 0.701$. Alternatively it can be determined from equation 3.2

$Pr = \mu C/k = [0.00001962 \times 1.0063]/(0.00002816)$

from which, $Pr = 0.702$.

As air flow is laminar and convection is free, equation 3.13 can be adopted thus:

$$Nu = 0.53(7.87 \times 10^6 \times 0.701)^{0.25} = 25.7$$

From equation 3.3,

$$h_c = Nu \cdot k/d = [25.7 \times 0.02816]/0.112 = 6.46 \text{ W/m}^2 \text{ K}.$$

From equation 3.16,

$$Q = 6.46 \times (\pi \times 0.112 \times 15)(79 - 19) = 2046 \text{ W}.$$

The calculation of heat radiation (equation 4.12 and $I = Q/A$) can be made from:

$$Q = \sigma \cdot e_1(T_1^4 - T_2^4)A_1$$

Substituting:

$$Q = 5.67 \times 10^{-8} \times 0.9[(352)^4 - (287)^4](\pi \times 0.112 \times 15)$$
$$= 2307 \text{ W}$$

The total output of the pipe coil $= 2046 + 2307 = 4353$ W.

3.6 Heat transfer from a tube in a condensing secondary fluid

Example 3.6 Determination of the convective heat transfer coefficient and surface area for a water cooled condenser

(a) Calculate the convective heat transfer coefficient at the inside surface of a 24 mm diameter tube in which pumped water is flowing at 0.5 kg/s, given water flow and return temperatures of 15°C and 25°C. Evaluate the properties at the mean bulk temperature.
(b) If ammonia vapour at a pressure of 15.54 bar is condensing on the outside surface of the tube which has an outside diameter of 30 mm, determine the surface area of the tube. Take the heat transfer coefficient at the outer surface of the tube for the condensing ammonia as 10 kW/m²K and ignore the temperature drop through the tube wall.

Solution (a)
To establish that the flow of water is turbulent, the Reynolds number can be used and from equation 3.1, $Re = dM/\mu A$. The mean bulk temperature of the

water flowing $= (15+25)/2 = 20°C$. From the tables for *Thermodynamic and Transport Properties of Fluids* the following properties of water are obtained:

$C = 4.183$ kJ/kgK, $\mu = 0.001002$ kg/ms, $k = 0.000603$ kW/mK, $Pr = 6.95$.

Adopting equation 3.1

$$Re = [0.024 \times 0.5 \times 4]/[0.001002\pi(0.024)^2]$$
$$= 26473$$

Since $Re > 2500$, the water flow inside the tube is in the turbulent region and as the water is pumped, equation 3.10 for forced convection can be adopted thus:

$$(Nu)_d = 0.023(Re)_d^{0.8}(Pr)^{0.33}$$

substitute:

$$Nu = 0.023(26473)^{0.8}(6.95)^{0.33}$$

from which $Nu = 151$. From equation 3.2, $Nu = h_c \cdot d/k$. Substituting:

$$h_c = [151 \times 0.000603]/0.024 = 3.794 \text{ kW/m}^2\text{K}.$$

Solution (b)
From equation 3.17, the overall U value for the heat exchange tube can be determined.

$$U = 1/((1/h_{si}) + R_f + (1/h_{so})), \text{ assuming negligible fouling resistance}$$

$R_f = $ zero

As the heat transfer coefficient for radiation eh_r is negligible at the inside surface of the tube, $h_{si} = h_c = 3.794$ kW/m² K and h_{so} is given here as 10 kW/m² K. Substituting into equation 3.17, $U = 1/[(1/3.794) + (1/10)] = 2.75$ kW/m²K.

From the tables for *Thermodynamic and Transport Properties of Fluids* ammonia at 15.54 bar absolute has a saturation temperature of 40°C. Adopting the heat balance to determine the net surface area of the heat exchange tubes:

heat gain to water = heat loss from ammonia.

66 Heat convection

Since the heat loss from ammonia $Q = U \cdot A \cdot dt$ (from equation 3.18),

$$M \cdot C \cdot dt \text{ (water)} = U \cdot A \cdot dt \text{ (ammonia)}$$

Substituting:

$$0.5 \times 4.183 \times (25 - 15) = 2.75 \times A \times (40 - 20)$$

From which, $A = 0.38$ m². Now tube length $L = A/\pi d = 0.38/(\pi \times 0.03) = 4.03$ m of heat exchange tubing.

SUMMARY FOR EXAMPLE 3.6
You would have noticed the temperature differential between ammonia and the bulk temperature of water. Here it is 20 K. The rate of heat transfer is largely influenced by this temperature differential. If the temperature differential between the primary and secondary fluids is below 15 K, heat transfer is poor.

The efficiency of heat exchange is also dependent upon the extent of the contact of the fluid with the outer surface of the heat exchanger. The contact factor identifies the extent of this contact which is among other things dependent upon the velocity of the fluid over the surface of the heat exchanger. Refer to Section 3.1.

3.7 Cooling flux from a chilled ceiling

Example 3.7 Determination of the cooling flux from a chilled beam ceiling

A chilled beam ceiling operates using chilled water at a mean bulk temperature of 8°C in a room held at an air temperature of 20°C by means of a system of displacement ventilation, and mean radiant temperature is 17°C. Determine the cooling flux in W/m² of the chilled ceiling surface which has an emissivity of 0.9.

Solution
The first step is to determine the Grashof number which will establish the type of air flow over the ceiling surface. The mean film temperature at the ceiling surface will be $= 0.5(20 + 8) = 14°C = 287$ K. Interpolating the properties of dry air at 287 K from the tables for *Thermodynamic and Transport Properties of Fluids*:

$$\rho = 1.233 \text{ kg/m}^3, \quad \mu = 0.00001783 \text{ kg/m s},$$
$$\beta = 1/T_m = 1/287 = 0.0035$$

Let $x = 1$, $dt = (20 - 8)$. Substitute into equation 3.4

$$Gr = [0.0035 \times (1.233)^2 \times 1^3 \times (20-8) \times 9.81]/(0.00001783)^2$$

from which $Gr = 1.98 \times 10^9$. For turbulent flow $1.4 \times 10^5 < (Gr) < 3 \times 10^{10}$ and therefore air flow over the ceiling surface is turbulent and free, and equation 3.9 applies and $h_c = 1.7(t_s - t_f)^{0.33}$. Substituting:

$$h_c = 1.7(20-8)^{0.33} = 3.86 \text{ W/m}^2\text{K}$$

the convective heat transfer,

$$Q = h_c \cdot A \cdot dt$$

Thus the cooling flux by free convection

$$Q/A = I = h_c \cdot dt = 3.86 \times (20-8)$$

from which $I = 46.32$ W/m² of chilled ceiling surface.

The cooling flux by heat radiation is obtained from equation 4.12 thus:

$$I = \sigma \cdot e_1(T_1^4 - T_2^4) \text{ W/m}^2$$
$$T_1 = 273 + 17 = 290 \text{ K} \quad \text{and} \quad T_2 = 273 + 8 = 281 \text{ K}$$

Substituting

$$I = 5.67 \times 10^{-8} \times 0.9(290^4 - 281^4)$$

from which $I = 42.77$ W/m² of chilled ceiling surface. The total cooling flux to the chilled ceiling $I = 46.32 + 42.77 = 89.1$ W/m².

SUMMARY FOR EXAMPLE 3.7

Dimension x was given a value of 1.0 in the solution for the Grashof number as the cooling flux is for an area of 1 m × 1 m of chilled ceiling surface. As the air flow over the ceiling is in the turbulent region, the final solution will not be affected if the cooling flux is now applied to a given area of chilled ceiling.

Had the air flow been in the laminar region, equation 3.7 would apply, in which case $D = (\text{length} + \text{width})/2$. This clearly affects the value of the heat transfer coefficient for convection h_c. The value of x in the determination of the Grashof number can again be taken as 1.0 without serious error.

3.8 Heat flux off a floor surface from an embedded pipe coil

Example 3.8 Calculation of the heat flux from the surface of a floor containing embedded pipe coils

(a) Determine the heat flux from the surface of a floor in which a hot water pipe coil is embedded in a room held at an air and mean radiant temperature of 20°C. The average floor temperature is 26°C and its emissivity is 0.9.
(b) Calculate the conductance h_s at the floor surface under the conditions in part (a).

Solution
(a) The Grashof number will be evaluated first to establish the type of air flow over the floor. The mean bulk temperature is $(26+20)/2 = 23°C = 296$ K. From the tables for *Thermodynamic and Transport Properties of Fluids*, the following properties of dry air at 296 K are obtained:

$$C = 1.0049 \text{ kJ/kgK}, \quad \mu = 0.00001846 \text{ kg/m s}, \quad \rho = 1.177 \text{ kg/m}^3$$
$$\beta = 1/T_m = 1/296 = 0.00338$$

Substitute into the Grashof formula equation 3.4

$$Gr = [0.00338 \times (1.177)^2 \times 1^3 \times (26-20) \times 9.81]/(0.00001846)^2$$

From which $Gr = 8.14 \times 10^8$. For turbulent flow, $3 \times 10^7 < (Gr) < 3 \times 10^{10}$. Thus, flow of air over the floor surface is in the turbulent region and free and equation 3.9 applies:

$$h_c = 1.7(26-20)^{0.33} = 3.071 \text{ W/m}^2\text{K}$$

The free convective heat flux, $I = h_c \cdot dt = 3.071 \times (26-20) = 18.42$ W/m². The heat radiation flux from the floor surface, using the appropriate equation 4.12,

$$I = \sigma \cdot e_1(T_1^4 - T_2^4) = 5.67 \times 10^{-8} \times 0.9[(299)^4 - (293)^4]$$
$$= 31.77 \text{ W/m}^2$$

The combined heat flux from the floor $I = 18.42 + 31.77 = 50.2$ W/m².
(b) By adapting equation 4.13 the heat transfer coefficient for radiation eh_r at the floor surface will be:

$$eh_r = I/dt = 31.77/(26-20) = 5.295 \text{ W/m}^2\text{K}$$

Heat convection 69

The heat transfer coefficient for free convection h_c at the floor surface is calculated as: $h_c = 3.071$ W/m²K. Since air and mean radiant temperature are equal, the coefficients can be combined and surface conductance h_s will be: $h_s = 5.295 + 3.071 = 8.366$ W/m²K.

SUMMARY FOR EXAMPLE 3.8
The solution to part (b) falls within the range of 8–10 W/m²K of published values for the combined heat transfer coefficient at the floor surface for embedded pipe coils. The maximum floor surface temperature should not exceed 27°C to avoid discomfort.

Example 3.9 Determining the length of a single pass shell and tube condenser

A single pass shell and tube condenser is required to condense refrigerant at 45°C. It contains 10 tubes each having an internal diameter of 24 mm and an external diameter of 30 mm. Cooling water is available at 10°C and total flow is 6.0 kg/s. Given that the temperature of the cooling water leaving the condenser is not to exceed 20°C and ignoring the temperature drop through the tube wall, determine the minimum tube length required. Take the heat transfer coefficient of the condensing vapour as 6.4 kW/m²K. The water side heat transfer coefficient should be determined from the appropriate rational equation.

Solution
Fluid flow in heat exchangers is discussed in some detail in Chapter 9. It can be assumed here that the water flow is subject to a prime mover and thus convection is forced. The Reynolds number Re can be calculated to establish turbulent flow conditions inside the tubes. From the tables for *Thermodynamic and Transport Properties of Fluids*, condenser water at a bulk temperature of $(20 + 10)/2 = 15°C = 288$ K has the following properties:

$$v = 0.001 \text{ m}^3/\text{kg}, \quad C = 4.186 \text{ kJ/kgK}, \quad \mu = 0.001136 \text{ kg/m s},$$
$$k = 0.595 \text{ W/mK}, \quad Pr = 7.99$$

From these properties $\rho = 1/0.001 = 1000$ kg/m³.

Now volume flow rate $= u \times \pi d^2/4 = Mv$ m³/s. Rearranging the formula for volume flow rate in terms of mean velocity, then mean velocity in each tube

$$u = (4 Mv/\pi d^2) = [4 \times (6.0/10) \times 0.001]/[\pi \times (0.024)^2]$$
$$= 1.326 \text{ m/s}$$

Substitute into equation 3.1

$$Re = [1000 \times 1.326 \times 0.024]/0.001136 = 28\,014$$

70 Heat convection

The minimum value for Re for turbulent flow inside tubes is 2500 and therefore flow is in the turbulent region here and equation 3.10 applies:

$$Nu = 0.023(28014)^{0.8}(7.99)^{0.33}$$

From which, $Nu = 165$. From equation 3.2, $h_c = (165 \times 0.595)/0.024 = 4091$ W/m²K.

The overall heat transfer coefficient U across the tubes is obtained from equation 3.17 and ignoring the effect of the fouling resistance R_f,

$$U = 1/[(1/6.4) + (1/4.09)] = 2.495 \text{ kW/m}^2\text{K}$$

The output from each tube in the condenser is calculated from $Q = M \cdot C \cdot dt$. Thus,

$$Q = (6.0/10) \times 4.2 \times (20 - 10) = 25.2 \text{ kW}$$

The overall heat transfer for each tube in the condenser is obtained from equation 3.18. Thus,

$$Q = U \cdot (\pi dL) \cdot dt = 2.495 \times (\pi \times 0.03 \times L) \times (45 - 15)$$

Using a heat balance in which:

heat gain by condenser water = heat loss by refrigerant.

Then by substitution:

$$25.2 = 2.495 \times \pi \times 0.03 \times L \times 30$$

from which tube length $L = 3.57$ m.

SUMMARY FOR EXAMPLE 3.9
No account has been taken here of the efficiency of heat exchange which is largely dependent upon the contact factor of the condensing refrigerant on the heat exchange tubes and upon the primary to secondary temperature difference which here is 30 K and well above the minimum of 15 K.

3.9 Heat transfer notes

The heat transfer coefficient

The heat transfer coefficients h_{si} and h_{so} relate to the transfer of heat across the fluid film at the inside and outside surface of a solid boundary. The film of fluid is known as the laminar sublayer and this is discussed in Chapter 6 with respect to fluid flow in pipes and ducts. The solid boundary in heat convection might be the heat exchanger tube. The solid boundary in heat conduction could be the external envelope of a building.

In this chapter on heat convection the thickness of the surface film in laminar flow is substantial and hence the magnitude of the heat transfer coefficient is relatively low and will have the effect of reducing the overall thermal transmittance of the heat exchanger. In turbulent flow the film thickness is reduced and hence the magnitude of the heat transfer coefficient is high.

This will increase the overall thermal transmittance of the heat exchanger and also its output. If the fluid film thicknesses at the inside and outside surface of the heat exchanger are both reduced due to turbulence the heat transfer coefficients will both be high in value and the overall thermal transmittance U will be at its maximum.

In heat conduction through an external wall, the film of air against the outside surface will be reduced in thickness as wind speed increases and h_{so} will therefore increase. This will have the effect of slightly increasing the thermal transmittance coefficient for the wall.

In heat conduction, the term surface conductance coefficient is used in place of the heat transfer coefficient. The following formulae and nomenclature relating to the surface film are used in this book.

Heat conduction: The combined surface conductance $h_{si} = (h_c + eh_r)$ W/m^{2}K and the combined surface conductance $h_{so} = (h_c + eh_r)$W/m^{2}K, with appropriate values adopted for h_c and eh_r. Thermal resistance $R_{si} = 1/h_{si}$ and $R_{so} = 1/h_{so}$ m^{2}K/W. This assumes that air and mean radiant temperature are equal. When they are assumed to be equal the term ambient temperature is used.

Heat convection: The heat transfer coefficient at the inside surface h_{si} is equal to the heat convection coefficient h_c. Likewise, the heat transfer coefficient at the outside surface h_{so} is equal to the appropriate heat convection coefficient h_c.

Heat radiation: The heat transfer coefficient at the surface for a black body is h_r W/m^{2}K, for a grey body it is eh_r and for a selective body it is $(F_1, F_2)h_r$.

3.10 Chapter closure

You now have the practical skills and the underpinning knowledge relating to the application of heat transfer by free and forced convection. The examples and solutions will have given some insight into the use of the rational formulae for this mode of heat transfer at steady temperatures and the procedures for problem solving. Further work on heat exchangers is done in Chapter 9. Recourse should be made to Appendix 2 for the origins of the dimensionless groups employed in this mode of heat transfer and to Chapter 6 for a detailed analysis of laminar and turbulent flow.

Chapter 4

Heat radiation

Nomenclature

a	absorptivity
A	surface area of radiator/receiver (m^2)
c	velocity of wave propagation (m/s)
C_1	constant (Wμm^4/m^2)
C_2, C_3	constants (μmK)
dt	temperature difference (K)
e	emissivity
eh_r $(F_{1,2})h_r$	heat transfer coefficient for radiation from a grey body (W/m^{2}K)
exp.	exponential e^x
$F_{1,2}$	form factor, view factor
h_r	heat transfer coefficient for radiation from a black body (W/m^{2}K)
I	heat flux, intensity of monochromatic radiation, intensity of heat radiation exchange solar constant (W/m^2)
ø	function of
prt	plane radiant temperature (°C)
Q	rate of energy flow (W)
r	reflectivity
S	distance between radiator and receiver (m)
T	absolute temperature (K)
t	transmissivity
t_a	air temperature (°C)
t_c	dry resultant temperature/comfort temperature (°C)
t_r	mean radiant temperature (°C)
t_s	surface temperature (°C)
U	thermal transmittance coefficient (W/m^{2}K)
v	frequency (Hertz, cycles/s)
vrt	vector radiant temperature (°C)
λ	wavelength (μm)
σ	Stefan–Boltzman constant of proportionality (W/m^{2}K^4)

4.1 Introduction

A simple definition of heat radiation would be: the interchange of electromagnetic waves between surfaces of differing temperatures which can see each other. In fact, the full definition is extensive and complex and requires substantial initial briefing and qualification before recourse can be made to practical applications.

All surfaces which are above absolute zero ($-273.15°C$ or 0.0 K) are emitting radiant heat or absorbing, reflecting and transmitting heat radiation depending upon whether they are emitting surfaces or receiving surfaces and upon whether the material is opaque or transparent. The distinction between an emitting surface and a receiving surface is dependent upon its temperature in relation to other surfaces it can 'see'.

One of the major differences between heat radiation exchange and that of heat conduction and heat convection is that it does not require an exchange medium. The sun transfers its heat by radiation through space to the Earth's atmosphere through which it passes to the surface of the Earth. The ozone layer, atmospheric particles, condensing water vapour and dust act as filters to solar radiation reducing its intensity at the Earth's surface. The effect of dust and other particles in the air has the same filtering effect for radiant space heaters which results in a reduction in heat flux at the receiving surfaces.

If the heat radiator cannot 'see' the surfaces to be heated, the effects of heat radiation are not immediately apparent if at all. Heat radiation is part of the spectrum of light ranging from the ultraviolet to the infrared region – that is, from short wave to long wave radiation. Space heating equipment which provides heat radiation include luminous heaters having temperatures up to 1200°C which emit short wave radiation and non-luminous heaters which emit invisible long wave radiation. Examples of each include the electric quartz heaters and radiant strip and, of course, the ubiquitous panel radiator both of which emit heat radiation which can be felt but not seen.

4.2 Surface characteristics

Heat radiation travels in the same wave patterns as light – see Figure 4.1 – and at the speed of light which is 2.98×10^8 m/s, frequently taken as $c = 3 \times 10^8$ m/s. The effectiveness of radiation exchange is dependent upon the texture of the radiating and receiving surfaces. Surface characteristics include: reflectivity (r), transmissivity (t), absorptivity (a) and emissivity (e).

If a mirror receives heat radiation, it will reflect about 97%, thus $r = 0.97$ and its ability to absorb radiation will be about 3%, thus $a = 0.03$. Its surface temperature therefore will not rise by much. Most receivers will reflect and absorb proportions of the incident radiation. A perfect radiator will emit 100% of its radiation received. The sun is a perfect radiator (known as

74 Heat radiation

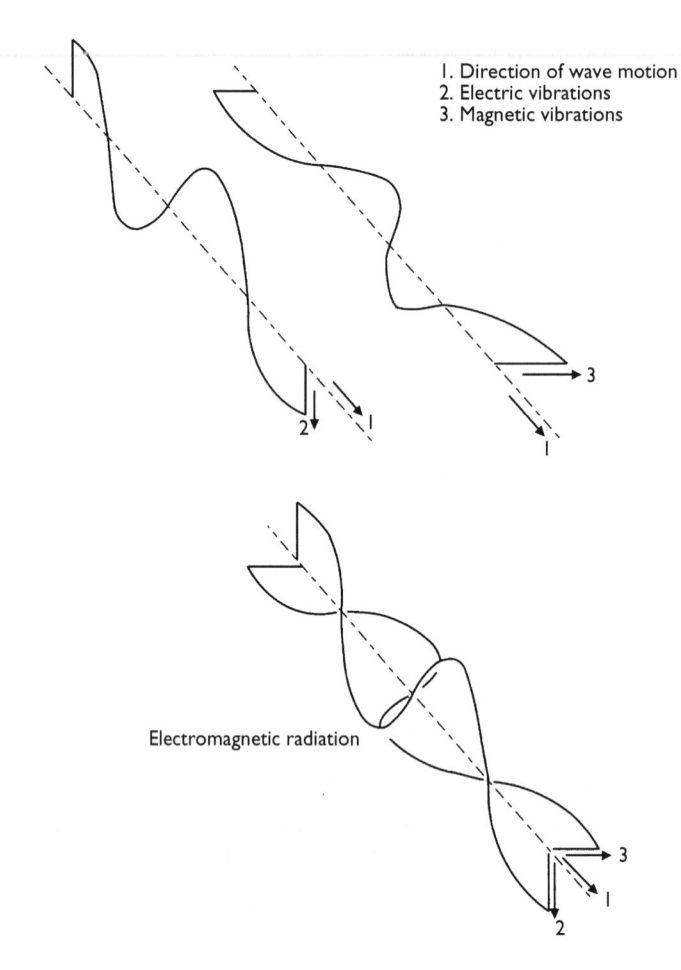

1. Direction of wave motion
2. Electric vibrations
3. Magnetic vibrations

Electromagnetic radiation

Figure 4.1 Transverse wave motion of electric and magnetic fields vibrating in phase at the same frequency at right angles in a plane perpendicular to the direction of travel.

'black body' radiation) but most surfaces are not and a radiator having a matt surface, for example, will emit about 90% of its radiation thus $e = 0.9$.

Kirchoff established that for most surfaces the ability to emit and absorb heat radiation at the same absolute temperature is approximately equal, thus, $e = a$ (equation 4.4). For most surfaces $r = (1 - a) = (1 - e)$ and therefore $a = (1 - r)$ and $e = (1 - r)$.

The transmission t of radiant heat through a material occurs by heat conduction resulting from the temperature rise induced by the radiant heat incident upon the surface. It is therefore similar to the thermal transmittance coefficient U. Refer to Figure 4.2. If a material which is transparent

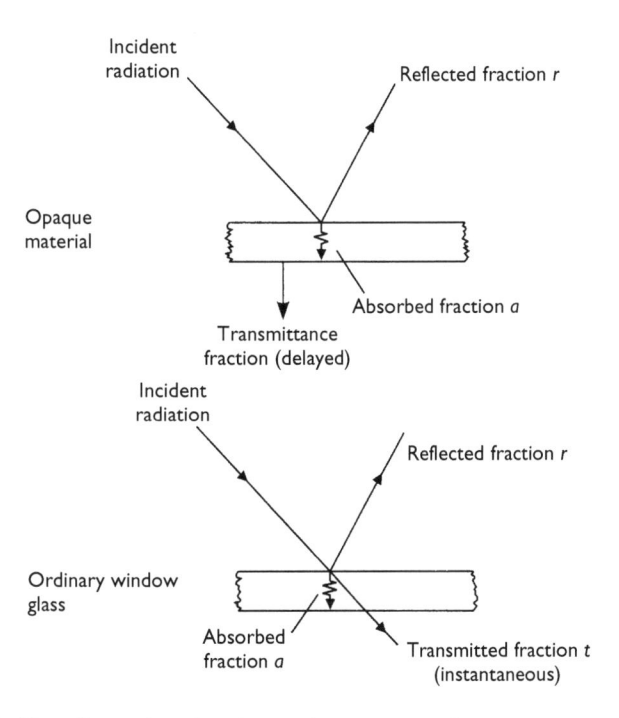

Figure 4.2 The effects of incident heat radiation on a surface.

Table 4.1 Surface characteristics relating to incident heat radiation

Material/surface	Emissivity (e)	Absorptivity (a)	Reflectivity (r)
Brick and stone	0.9	0.9	0.1
Aluminium, polished	0.04	0.04	0.96
Aluminium, anodised	0.72	0.72	0.28
Cast iron	0.8	0.8	0.2
Copper, polished	0.03	0.03	0.97
Copper, oxidised	0.86	0.86	0.14
Galvanised steel	0.25	0.25	0.75
Paint, metal based	0.5	0.5	0.5
Paint, gloss white	0.95	0.95	0.05
Paint, mattt black	0.96	0.96	0.04

like ordinary window glass is irradiated, the short wave radiation will be transmitted with little absorption or reflection.

Table 4.1 gives some typical values for emissivity, absorptivity and reflectivity for some opaque surfaces. The table gives a general indication only since for many materials emissivity varies with temperature.

76 Heat radiation

Table 4.2 Colour temperature guide for luminous heaters

Colour	Temperature (°C)
Very dull red	500–600
Dark blood red	600–700
Cherry red	700–800
Bright red	800–900
Orange	900–1000
Yellow	1000–1100
Yellow/white	1100–1200
White	1200–1300

Colour indicators

It is useful to have an approximate feel for the temperature of luminous radiant heaters.

Table 4.2 gives a rough guide.

4.3 The greenhouse effect

Greenhouse glass allows the transmission of short wave solar radiation but disallows long wave transmission. The resultant effect is known as the greenhouse effect. As the surfaces within the greenhouse warm up due to the incidence of short wave solar radiation passing through the glass, they begin to emit long wave radiation which cannot escape resulting in a rise in temperature within the greenhouse. This is used to good effect for the propagation of plants and as a means of passive space heating in the winter.

4.4 Spectral wave forms

The pattern of solar radiation follows a sine wave, see Figure 4.3. The spectral proportions in which heat radiation is present are given in Table 4.3.

A discussion on heat radiation needs to identify three types of emitting and receiving opaque surfaces in order to reduce the complexity of the subject. These are black, grey and selective. A 'black' surface is that of a perfect radiator in which emissivity is constant at any wavelength. Black body (perfect) radiation is rare. It is approached within a boiler furnace and with luminous radiant heaters but not often elsewhere in the real world. Selective surfaces are those of every day manufactured or natural materials. The emissivity of selective surfaces varies arbitrarily with wavelength. This makes it difficult to integrate the radiant heat flux over all the wavelengths for a given temperature of a selective emitter or receiver. A grey surface is an imaginary surface in which the emissivity varies uniformly with wavelength making it

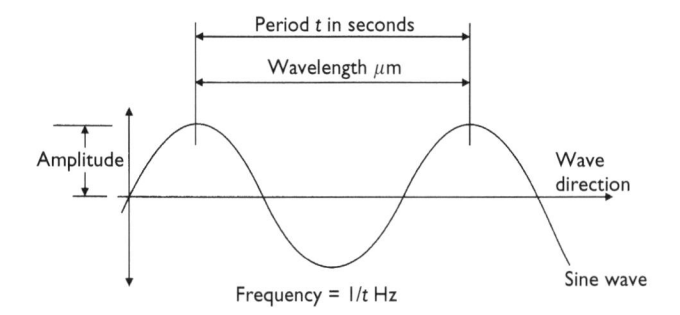

Figure 4.3 Heat radiation waveform.

Table 4.3 Spectral proportions of heat radiation

Cosmic, gamma and X ray wavebands, wavelength	10^{-11}–10^{-2} μm
9% ultraviolet, invisible short wave, wavelength	0.29–0.4 μm
40% visible light, short wave, wavelength	0.40–0.7 μm
51% invisible infrared, long wave, wavelength	0.70–3.5 μm
Radio, TV and radar wavebands, wavelength	100 μm–10^5 m

easier to integrate the radiant heat flux over all the wavelengths for the given surface. Since the grey surface has a constant heat flux ratio with that of a black body, $I_\lambda/I_{b\lambda}$, it irons out the arbitrary nature of the selective surface making it easier to determine the heat flux I.

4.5 Monochromatic heat radiation

Heat radiation emitted at any one wavelength is called monochromatic radiation. Figure 4.4 shows the variation in black body emissive power with wavelength and absolute temperature for two bodies at temperature T_1 and T_2. Note how wavelength increases as maximum absolute temperature of the body decreases. This was identified by Wein and is known as Wein's displacement law. See equation 4.5. It is denoted by a uniform line declining towards the right on the graph.

The area under each of the curves at T_1 and T_2 represents the sum of the monochromatic radiations or total heat radiation from the surfaces. A comparison of emissions from black, grey and selective surfaces is shown in Figure 4.5. Note that the ratio of $I_\lambda/I_{b\lambda}$ is constant for the grey surface and arbitrary for the selective surface where it varies at each wavelength. Note also that the absolute temperature of each surface is the same. It is the monochromatic emissive power from each surface which varies with wavelength.

78 Heat radiation

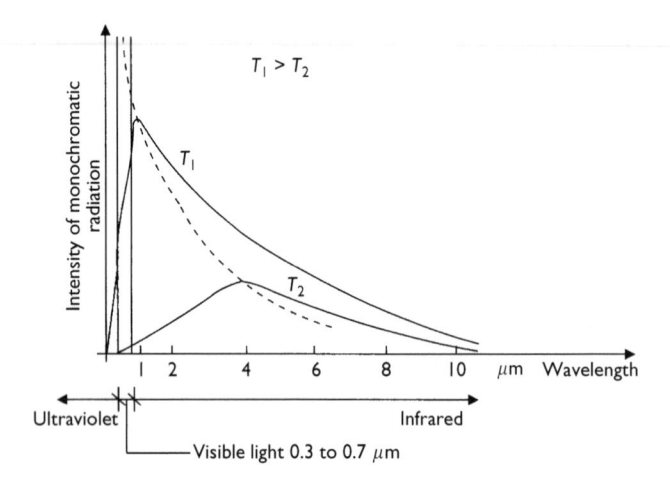

Figure 4.4 Variation of black body emissive power with wavelength and absolute temperature.

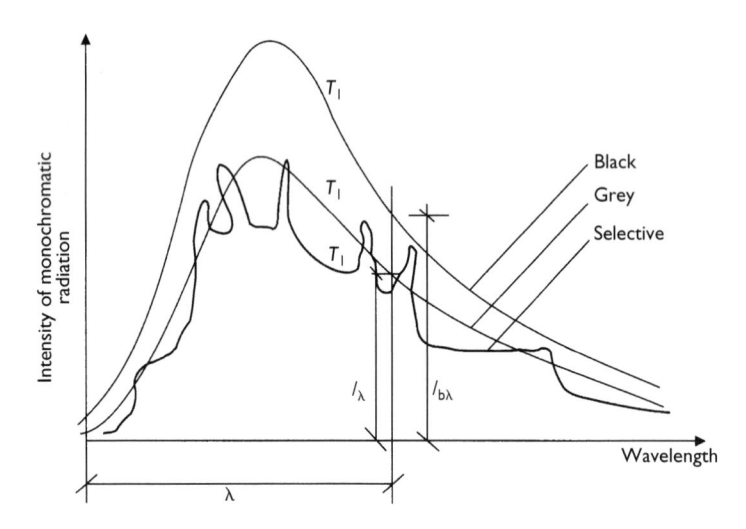

Figure 4.5 Comparison of emission from black, grey and selective surfaces at a fixed temperature T_1.

4.6 Laws of black body radiation

The following laws apply to heat radiation from a perfect radiator:

Kirchoff's law and Stefan's law – heat flux, $I \propto T^4$ W/m²   (4.1)

Heat radiation

The Stefan–Boltzman constant of proportionality,

$$\sigma = 5.67 \times 10^{-8} \text{ W/m}^2\text{K}^4$$

Thus, heat radiation for sum of the wavelengths $I = \sigma \cdot T^4$ W/m²

(4.2)

Planck's law shows the relationship between the monochromatic emissive power I at wavelength λ and absolute temperature T, thus

$$I = C_1 \cdot \lambda^{-5}/[\exp(C_2/\lambda T) - 1] \text{ W/m}^2 \tag{4.3}$$

Kirchoff's law, $e = a$ (4.4)

Wein's displacement law, $\lambda_{\max} \cdot T = C_3 \ \mu\text{m K}$ (4.5)

Wein's wavelength law, $\lambda = c/v$ m (4.6)

Lambert's law for emissive power from a flat radiator $I_\phi = \ln \cos \phi$

(4.7)

See Figure 4.6.

Stefan–Boltzman law for heat radiation exchange over the sum of the wavelengths at temperature T_1,

$$I = \sigma(T_1^4 - T_2^4) \text{ W/m}^2 \tag{4.8}$$

The constants above have the following numerical values:

$c = $ velocity of wave propagation $= 3 \times 10^8$ m/s

$C_1 = 3.743 \times 10^8 \text{ W}\mu\text{m}^4/\text{m}^2$

$C_2 = 1.4387 \times 10^4 \ \mu\text{m K}$

$C_3 = 2897.6 \ \mu\text{m K}$

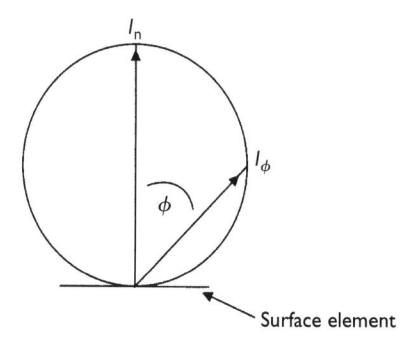

Figure 4.6 A surface element does not radiate energy with equal intensity in all directions, $I_\phi = \ln \cos \phi$.

4.7 Laws of grey body radiation

When radiating surfaces remain grey in a system of heat exchange, the emissivities of those surfaces must be accounted for, as well as their geometric configuration. In general, the heat exchange by radiation between two surfaces will depend upon

- relative areas of surfaces;
- geometry of the surfaces in relation to each other;
- the two emissivities.

These factors which are quite complex are identified as the form/view factor $F_{1,2}$.

$$F_{1,2} = \emptyset(A_1, A_2, e_1, e_2)$$

The view factor was introduced in 1951 by H.C. Hottell as a means of accounting for the emissivities of the surfaces and their geometric configuration. The *CIBSE Guide* tabulates form factors for various surface configurations. In view of the complexity of determining the form factor, only three applications will be considered here.

1. *Two parallel grey surfaces in which* $A_1 = A_2$ *and* $T_1 > T_2$. A typical application here is the radiation exchange between the two inside surfaces of a cavity wall.

$$F_{1,2} = 1/[(1/e_1) + (1/e_2) - 1] \quad (4.9)$$

2. *Concentric cylindrical surfaces in which* $A_1 < A_2$ *and* $T_1 > T_2$. A typical application is layers of thermal insulation on cylindrical ducts and pipes.

$$1/F_{1,2} = (1/e_1) + (A_1/A_2)((1/e_2) - 1) \quad (4.10)$$

3. *A small radiator contained in a large enclosure in which* $A_1 < A_2$ *and* $T_1 > T_2$.

$$F_{1,2} = e_1$$

This relationship is approximately correct for most applications of space heating and cooling in which A_2 is the surface area of the enclosing space and T_2 its area weighted mean surface temperature or mean radiant temperature.

Thus for grey body heat radiation $I = (F_{1,2})\sigma(T_1^4 - T_2^4)$ W/m^2 (4.11)

4.8 Radiation exchange between a grey body and a grey enclosure

Clearly an imaginary surface whose characteristics iron out the variations in emissivity at different wavelengths and which follows a constant heat flux ratio is more easily analysed than the random variations in emissivity with wavelength which occurs with a selective surface.

Consider heat radiation exchange between a small grey body radiator located in a grey enclosure as shown in Figure 4.7. Assume that the absorptivity of the enclosure is 0.9 and that the radiator emits 100 units of heat radiation. Of the 100 units, 99.9 are absorbed by the enclosure, so the apparent absorptivity of the surrounding enclosure boundaries is 0.999 Although this is a simplified analysis of the matter, it can be argued that the boundary surfaces, initially taken as grey, approach that of a black body enclosure.

Thus, from equation 4.11 and the notes in Section 4.7,

$$I = \sigma \cdot e_1(T_1^4 - T_2^4) \text{ W/m}^2 \tag{4.12}$$

where T_1 is the absolute temperature of the radiator and T_2 is the absolute area weighted mean radiant temperature of the surrounding surfaces which is approximately equal to the mean radiant temperature of the enclosure T_r.

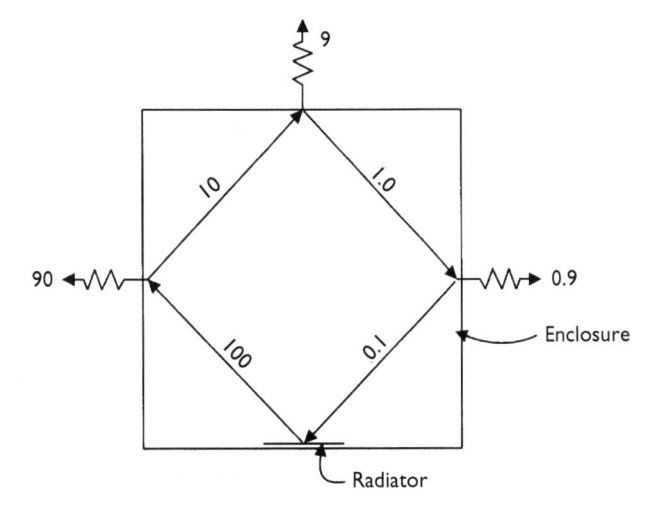

Figure 4.7 Heat radiation exchange. If the enclosure has an absorptivity of 0.9 then out of 100 units of radiant heat 99.9 are absorbed.

4.9 Heat transfer coefficients for black and grey body radiation

The heat transfer coefficient for convection h_c, Chapter 3, equation 3.16, is:

$$Q = h_c \cdot A \cdot dt \text{ W}$$

Thus for heat convection

$$I = h_c(t_1 - t_2) = h_c(T_1 - T_2) \text{ W/m}^2$$

For black body heat radiation exchange,

$$I = h_r(T_1 - T_2) \text{ W/m}^2$$

Thus

$$h_r = I/dT \text{ W/m}^2 \text{ K} \tag{4.13}$$

The heat transfer coefficient for radiation can also be determined from above, where $I = h_r(T_1 - T_2)$ W/m^2 and from equation 4.8, $I = \sigma(T_1^4 - T_2^4)$ W/m^2. Equating these two formulae

$$h_r = \sigma(T_1^4 - T_2^4)(T_1 - T_2)^{-1}$$

Expanding

$$h_r = \sigma(T_1^2 - T_2^2)(T_1^2 + T_2^2)(T_1 - T_2)^{-1}$$

and

$$h_r = \sigma(T_1 - T_2)(T_1 + T_2)(T_1^2 + T_2^2)(T_1 - T_2)^{-1}$$

From which for black bodies

$$h_r = \sigma(T_1 + T_2)(T_1^2 + T_2^2) \text{ W/m}^2\text{K} \tag{4.14}$$

For grey bodies

$$(F_{1,2})h_r = \sigma(F_{1,2})(T_1 + T_2)(T_1^2 + T_2^2) \tag{4.15}$$

For a small radiator in a large enclosure,

$$(e_1)h_r = \sigma(e_1)(T_1 + T_2)(T_1^2 + T_2^2) \tag{4.16}$$

Note that in equations 4.15 and 4.16 $(F_{1,2})$ and (e_1) do not cancel in the determination of the heat transfer coefficient for radiation for grey bodies.

4.10 Heat radiation flux I

As the distance S between the receiving surface and the emitting surface increases, the less intense is the radiation flux. This is borne out by varying the distance a person is with respect to a radiant heater like a luminous electric fire. Three applications are considered here.

1. *Point source radiation*: The direction of intensity is spherical here hence A_2 is the surface area of a sphere and $A_2 = 4\pi S^2$. If the radius is doubled – that is, to say if the distance S from the point source radiator is doubled, the enclosing receiving area A_2 is quadrupled.

    Thus if originally the distance between the point source radiator and the enclosing area is 3, $S = 3$ and $A_2 = 4\pi 3^2 = 36\pi$; whereas if S is doubled to 6, $A_2 = 4\pi 6^2 = 144\pi$, which is four times larger.

    The effect upon radiation flux or intensity is a four-fold reduction. The total heat radiation received by the enclosing surface of radius 3 is the same as that for the enclosing surface of radius 6, however. Thus

    $$I \propto 1/S^2$$

2. *Line source radiator*: Here $I = 1/S$. An example might be a single ceramic luminous rod heater with no reflector and considering radiant heat flux in one plane only.

3. *A surface element*: It does not radiate energy with equal intensity in all directions and Lambert's law applies, equation 4.7. See Figure 4.6. It identifies the fact that the greatest radiant heat flux received occurs along a line normal to the radiating surface. If it is assumed that the lines of heat radiation from a flat panel radiator are not parallel but expanding as shown in Figure 4.8 then the total heat radiation received at location 2 will be the same as at location 3.

    Thus the products of $I_1 \cdot A_1 = I_2 \cdot A_2 = I_3 \cdot A_3$ Watts, where radiant heat flux I is taken as a mean value at both locations. The heat flux at each location gets progressively less in direct proportion to the increase in receiving areas.

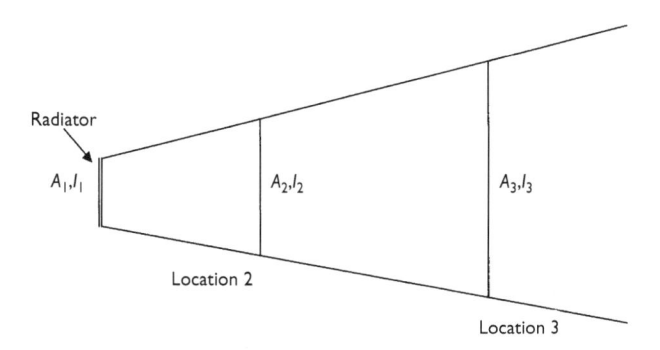

Figure 4.8 Heat radiation received at locations 2 and 3.

4.11 Problem solving

The foregoing discussion provides an introduction to the subject of heat radiation.

It should, however, be sufficient to form the underpinning knowledge for the building services engineer. There now follows some examples involving this mode of heat transfer.

Example 4.1 Calcuation of the radiant emission from a luminous quartz heater

A luminous quartz heater has a temperature of 1200°C and an effective area of 0.3 × 0.3 m Determine (a) the total rate of radiant emission; (b) the wavelength of maximum energy; (c) the monochromatic emissive power at the wavelength of maximum energy.

Solution
You will notice that there is no indication of the mean radiant temperature of the surfaces in the enclosure in which the heater is located so the solution will not account for radiation exchange. Since the heater is luminous it is assumed to be a black body,
(a) Thus from equation 4.2

$$I = 5.67 \times 10^{-8}(1200 + 273)^4$$

from which

$$I = 2.67 \times 10^5 \text{ W/m}^2 = 267 \text{ kW/m}^2$$

This is the radiant heat flux for the sum of all the wavelengths. The effective area is given as $A = 0.3 \times 0.3 = 0.09$ m^2. Therefore, total radiant emission $= 267 \times 0.09 = 24$ kW.
(b) From equation 4.5, $\lambda_{max} \cdot T = C_3$, where $T = 1473$ K. Thus $\lambda_{max} = C_3/T = 2897.6/1473 = 1.97$ μm.
(c) From equation 4.3,

$$I = (3.743 \times 10^8 \times 1.97^{-5})/(\exp(1.4387 \times 10^4/1.97 \times 1473) - 1)$$

Given that $\exp = e^x = 2.7183^x$, where $x = (1.4387 \times 10^4/1.97 \times 1473)$.

$$I = (3.743 \times 10^8 \times 1/29.67)/[(2.7183^{4.958}) - 1]$$

$$I = [1.2615 \times 10^7]/(142 - 1)$$

$$I = 8.95 \times 10^4 = 89.5 \text{ kW/m}^2$$

This is the emissive power at the wavelength of maximum heat radiation.

For the effective area of the radiator, the emissive power at this wavelength = 89.5 × 0.09 = 8.055 kW.

SUMMARY FOR EXAMPLE 4.1
The proportion of the emissive power of 89.5 kW/m² attributed at the wavelength of maximum heat radiation (1.97 μm), to the total emissive power across all wavelengths of 267 kW/m² is 89.5/267 = 33.5%. Thus one-third of the luminous heater output is derived from the wavelength of 1.97 μm.

Example 4.2 Determination of the effect of a bright aluminium foil located in an external cavity wall

Show the effect on radiant heat flux of locating bright aluminium foil having an emissivity of 0.07 in the centre of an external wall cavity. Assume that the two boundaries to the cavity are at 10°C and 1°C, respectively, each having an emissivity of 0.9. Comment upon the effect that the foil will have on the thermal transmittance coefficient for the wall and upon its thermal response.

Solution
Figure 4.9 shows the wall cavity with the foil in place. Absolute temperatures are used in the solution and

$$T_1 = (273 + 10) = 283 \text{ K}, \qquad T_2 = (273 + 1) = 274 \text{ K}$$

Equations 4.9 and 4.11 can be adopted here in which T_f is the absolute temperature of the foil. Thus if temperatures remain steady, heat flow from T_1 to T_f equals the heat flow from T_f to T_2 and

$$\sigma(T_1^4 - T_f^4)/((1/e_1) + (1/e_f) - 1) = \sigma(T_f^4 - T_2^4)/((1/e_f) + (1/e_2) - 1)$$

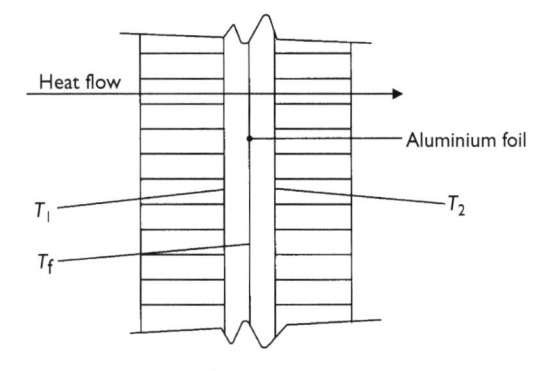

Figure 4.9 External wall cavity with bright aluminium foil in place (Example 4.2).

86 Heat radiation

Since the form factor has the same numerical value on each side of the heat balance and the Stefan–Boltzman constant cancels

$$(T_1^4 - T_f^4) = (T_f^4 - T_2^4)$$

and substituting:

$$283^4 - T_f^4 = T_f^4 - 274^4$$
$$283^4 + 274^4 = 2T_f^4$$

Thus

$$(6.414 \times 10^9) + (5.636 \times 10^9) = 2T_f^4$$

So

$$(1.2051 \times 10^{10}) = 2T_f^4$$

and

$$\sqrt[4]{\{(1.2051 \times 10^{10})/2\}} = T_f$$

from which foil temperature, $T_f = 278.6$ K ($t_f = 5.6°C$).

The radiant heat flux I for two parallel grey surfaces can be determined by combining equations 4.9 and 4.11,

$$I = (F_{1,2})\sigma(T_1^4 - T_2^4) \text{ W/m}^2$$

thus:

$$I = \sigma(T_1^4 - T_2^4)/((1/e_1) + (1/e_2) - 1)$$

The radiant heat flux between surfaces 1 and 2 before the foil is located will be:

$$I = [5.67 \times 10^{-8}(283^4 - 274^4)]/((1/0.9) + (1/0.9) - 1)$$

and

$$I = 44.1/1.222 = 36.1 \text{ W/m}^2$$

The radiant heat flux when the foil is in position can be determined from either one side or the other of the foil since the heat flux from surface 1 to the foil will equal the heat flux from the foil to surface 2 assuming temperatures remain steady. Taking then the heat flux from surface 1 to the foil:

$$I = [5.67 \times 10^{-8}(283^4 - 278^4)]/((1/0.9) + (1/0.07) - 1)$$

from which $I = 25.03/14.4 = 1.74$ W/m^2.

Heat radiation 87

The heat transfer coefficient for radiation $(F_{1,2})h_r$ can be determined from equation 4.15 and for parallel boundaries equation 4.9 is applicable for form factor $F_{1,2}$ and therefore when the foil is in position, working from boundary 1 to the foil:

$$(F_{1,f})h_r = \sigma(T_1 + T_f)(T_1^2 + T_f^2)/((1/e_1) + (1/e_f) - 1)$$

$$(F_{1,f})h_r = [5.67 \times 10^{-8}(283 + 278)(283^2 + 278^2)]/[(1/0.9) + (1/0.07) - 1]$$

$$(F_{1,f})h_r = 0.348 \text{ W/m}^2\text{K}$$

A much simpler way of finding $(F_{1,f})h_r$ here, since the radiant heat flux I has been calculated, is to use equation 4.13. Thus $(F_{1,f})h_r = I/dt = 1.74/(283 - 278) = 0.348$ W/m²K.

The heat transfer coefficient for radiation when the foil is not in place will take place between the two surfaces 1 and 2, thus using again equations 4.9 and 4.15:

$$(F_{1,2})h_r = [5.67 \times 10^{-8}(283 + 274)(283^2 + 274^2)]/[(1/0.9) + (1/0.9) - 1]$$

$$(F_{1,2})h_r = 4.01 \text{ W/m}^2\text{K}$$

Similarly again equation 4.13 can be used here and

$$(F_{1,2})h_r = I/dt = 36.1/(283 - 274) = 4.01 \text{ W/m}^2\text{K}$$

SUMMARY FOR EXAMPLE 4.2

Condition	Radiant heat flux I (W/m²)	Radiant heat transfer coefficient $(F_{1,2})h_r$ (W/m²K)
No foil	36.1	4.01
With foil	1.74	0.348

With the foil in place in the wall cavity the radiant heat flux is reduced to about 5% of its original value. If the foil is 2 mm thick and is made of aluminium, it will have a thermal conductivity of 105 W/mK. Its thermal resistance $R = 0.002/105 = 0.000019$ m² K/W. This will have no significant effect upon the thermal transmittance coefficient (U value) for the wall and therefore no effect upon the heat loss. However, with the foil reflecting back much of the radiant component of heat transfer towards the inner leaf of the wall, the inner leaf will absorb heat more quickly when the heating plant is started after a shut down period. The foil will also assist in

88 Heat radiation

maintaining the temperature of the inner leaf and hence indoor temperature longer when the plant shuts down at the end of the day. This results in energy conservation.

Example 4.3a Determination of the effect of a bright metal foil located behind a radiator

(i) A vertical panel radiator, fixed to an external wall, measures 1.8 by 0.75 m high and has a mean surface temperature of 76°C and an emissivity of 0.92. It is intended to fix a bright metal foil having an emissivity of 0.04 directly to the wall behind the radiator having an emissivity of 0.9. The room is held at a mean radiant temperature of 19°C and the wall temperature behind the radiator stabilises at 40°C. Determine the radiant heat emission from the panel before and after the foil is in place.

(ii) If the room air temperature is 21°C determine the heat output by free convection from the radiator. Evaluate the properties at the mean film temperature and assume one of the following relationships for the surface convection coefficient h_c. For turbulent flow $Nu = 0.1(Pr \cdot Gr)^{0.33}$, for laminar flow, $Nu = 0.36(Gr)^{0.25}$.

Solution 4.3(a)(i)
Figure 4.10 shows the radiator panel fixed to the external wall. Absolute temperatures are: panel surface, $76 + 273 = 349$ K; wall, foil surface, $40 + 273 = 313$ K; room, $19 + 273 = 292$ K. Radiant heat flow from the back of the panel must account for the form factor for parallel surfaces, equation 4.9. Radiant heat flow from the front of the panel involves the form factor $F_{1,2} = e_1$. Two separate calculations therefore need to be considered here for each condition.

Considering the back of the panel with no foil in place
From equation 4.9

$$F_{1,2} = 1/((1/e_1) + (1/e_2) - 1)$$

substituting $e_1 = 0.92$ and $e_2 = 0.9$,

$$F_{1,2} = 1/1.198 = 0.835$$

From equation 4.11 in which $T_1 = 349$ K and $T_2 = 313$ K, $I = 248$ W/m² and emission $Q = 248 \times 1.8 \times 0.75 = 335$ W.

Considering the front of the panel
From equation 4.12 in which $e_1 = 0.92$, $T_1 = 349$ K and $T_2 = 292$ K, $I = 395$ W/m² and emission $Q = 395 \times 1.8 \times 0.75 = 533$ W. Total radiant output from the panel without the foil $= 335 + 533 = 868$ W.

You should now confirm these calculations.

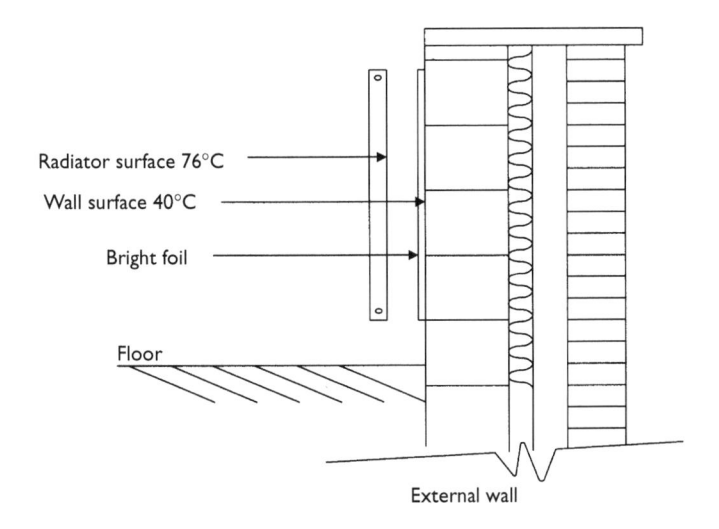

Figure 4.10 Use for bright reflective foil (Example 4.3).

Considering the back of the panel with the foil in place
From equation 4.9

$$F_{1,2} = 1/((1/0.92) + (1/0.04) - 1) = 1/25.087 = 0.04$$

From equation 4.11

$$I = 5.67 \times 10^{-8} \times 0.04[(349^4) - (313^4)] = 11.88 \text{ W/m}^2$$

and emission

$$Q = 11.88 \times 1.8 \times 0.75 = 16 \text{ W}$$

The front of the panel will have the same radiant emission as before and $Q = 533$ W.

The total radiant output from the panel with the foil in place = $533 + 16 = 549$ W.

SUMMARY FOR EXAMPLE 4.3 (a)(i)
With the foil in place, the radiant heat output from the panel is 549 W compared with 868 W. The heat loss by radiation through the wall is $868 - 549 = 319$ W when the foil is missing. There is therefore a saving in heat energy, and hence fuel costs, when the foil is used. There will also be a reduction in CO_2 emission into the atmosphere.

Solution 4.3(a)(ii)
From the chapter on heat convection, the first step in the solution is to establish whether the free convection over the panel is laminar or turbulent. This is achieved by evaluating the Grashof number,

$$Gr = \beta \cdot g \cdot x^3 \cdot \rho^2 \cdot dt/\mu^2$$

The mean film temperature at the panel surface $= (76+21)/2 = 48.5°C = 321.5$ K.

The density and viscosity of the air at the mean film temperature is interpolated from the tables of *Thermodynamic and Transport Properties of Fluids* as $\rho = 1.099$ kg/m^3, $\mu = 0.00001946$ kg/ms, $P_r = 0.701$ and $k = 0.02789$ W/mK. The coefficient of cubical expansion of the air $\beta = 1/321.5 = 0.00311$ K^{-1}.

Substituting these values into the Grashof formula:

$$Gr = [(0.00311 \times 9.81 \times (0.75)^3 \times (1.099)^2 (76-21)]/(0.00001946)^2$$
$$= 2.27 \times 10^9$$

Since $Gr > 10^9$ air flow over the panel is turbulent, thus

$$Nu = 0.1(0.701 \times 2.27 \times 10^9)^{0.33} = 108.8$$

Now $Nu = h_c \cdot x/k$ from which

$$h_c = Nu \cdot k/x = 108.8 \times 0.02789/0.75 = 4.04 \text{ W/m}^2\text{K}$$

Free convective emission

$$Q = h_c \cdot A \cdot dt = 4.04 \times (1.8 \times 0.75 \times 2)(76-21) = 600 \text{ W}.$$

COMPARISON OF SOLUTIONS 4.3(a)(i) WITH 4.3(a)(ii)
The total output from the panel with the foil in place $= 549 + 600 = 1149$ W.
The total output from the panel without the foil $= 868 + 600 = 1468$ W.

Example 4.3(b) Calculation of the upward and downward emission from a radiant panel

A horizontal radiant panel located at high level in a workshop and insulated on its upper face has a surface temperature of 110°C, an emissivity of 0.9 and measures 2.7 × 1.2 m. The workshop is held at a mean radiant temperature of 19°C and an air temperature of 15°C.

(i) Determine the total emission downwards.
(ii) Determine the total emission upwards if the outer surface of the insulation has a temperature of 30°C and an emissivity of 0.1.
(iii) Calculate the thickness of thermal insulation given that its thermal conductivity is 0.07 W/mK.

Data: Rational formulae:

$$I = (F_{1,2})\sigma[T_1^4 - T_2^4] \text{ W/m}^2 \tag{4.11}$$

$$\text{Downward } h_c = 0.64[(t_s - t_f)/D]^{0.25} \text{ W/m}^2 \text{ K} \tag{3.8}$$

$$\text{Upward } h_c = 1.7(t_s - t_f)^{0.33} \text{ W/m}^2 \text{ K} \tag{3.9}$$

$$Q = h_c \cdot A \cdot dt \text{ W} \tag{3.16}$$

Solution 4.3(b)(i)
Now

$$I = 5.67 \times 10^{-8} \times 0.9[(383)^4 - (292)^4]$$

from which $I = 727$ W/m², and $Q_r = 727(2.7 \times 1.2) = 2356$ W.
Now from equation 3.8,

$$h_c = 0.64[(110 - 15)/((2.7 + 1.2)/2)]^{0.25}$$

from which $h_c = 1.69$ W/m² K; and from equation 3.16

$$Q_c = 1.69(2.7 \times 1.2)(110 - 15)$$

from which $Q_c = 520$ W. Total downward emission $Q = 2356 + 520 = 2876$ W.

Solution 4.3(b)(ii)
The upper surface of the thermal insulation is finished in aluminium foil, hence the low value for its emissivity. Now

$$I = 5.67 \times 10^{-8} \times 0.1[(303)^4 - (292)^4]$$

from which $I = 6.57$ W/m² and $Q_r = 6.57(2.7 \times 1.2) = 21.3$ W.
Now from equation 3.9,

$$h_c = 1.7(30 - 15)^{0.33}$$

from which $h_c = 4.155$ W/m² K; and from equation 3.16

$$Q_c = 4.155(2.7 \times 1.2)(30 - 15)$$

from which $Q_c = 202$ W. Total upward emission $Q = 21 + 202 = 223$ W.

92 Heat radiation

Solution 4.3(b)(iii)
From equation 2.3,

$$dt_1/R_1 = dt_2/R_2 = I \text{ W/m}^2$$

Adapting the equation for use here where $t_1 = 110°C$ and $t_2 = 30°C$

$$I = (t_1 - t_2)/R_{\text{ins}} \text{ W/m}^2$$

Substituting:

$$223/(2.7 \times 1.2) = (110 - 30)/(L/0.07)$$

Thus

$$68.83 = 80/(L/0.07)$$

Rearranging:

$$68.83 \times L/0.07 = 80$$

from which $L = 0.081$ m $= 81$ mm.

SUMMARY FOR EXAMPLE 4.3(b)
Total downward emission = 2876 W, total upward emission = 223 W, thickness of insulation on the upper side of the panel = 81 mm.

Example 4.4 Calculation of the effects on heat emission from a steam pipe having different thermal insulation applications

(a) A steam pipe at a temperature of 200°C passes through a room in which the mean radiant temperature is 20°C. A short section of the pipe surface is uninsulated and its emissivity is 0.95. If its area is 0.25 m², calculate the rate at which heat radiation will be lost.
(b) How would heat emission be affected by:

   (i) Painting the pipe with aluminium paint given $e = 0.7$
   (ii) Wrapping the pipe tightly with aluminium foil given $e = 0.2$
   (iii) Surrounding the pipe with a co-axial cylinder of aluminium foil where its outside diameter is twice the diameter of the steam pipe.

Solution
Absolute temperatures are used and $T_1 = 273 + 200 = 473$ K, $T_2 = 273 + 20 = 293$ K.

(a) Assuming that the uninsulated pipe surface is small compared with the size of the room in which it is located equation 4.12 may be adopted in which $F_{1,2} = e_1$.
   Thus $I = 5.67 \times 10^{-8} \times 0.95(473^4 - 293^4)$
   and $I = 2299$ W/m²
   $Q = 2299 \times 0.25 = 575$ W

(b) (i) As the only variable is emissivity

$$Q = 575 \times \text{ ratio of emissivities} = 575 \times 0.7/0.95 = 425 \text{ W}$$

(ii) $\quad Q = 575 \times 0.2/0.95 = 120 \text{ W}$

(iii) The surface area of the aluminium casing at absolute temperature T_s will be twice that of the pipe. Thus the radiant heat loss from the casing surface to the room will be:

$$Q = 0.2\sigma(T_s^4 - 293^4)(2 \times 0.25)$$

The radiant heat transfer into the aluminium casing from the pipe surface from equations 4.10 and 4.11 will be:

$$I = \sigma(473^4 - T_s^4)/[(1/e_1) + A_1/A_2((1/e_2) - 1)] \text{ W/m}^2$$

Accounting for the surface area of the pipe:

$$Q = \sigma(473^4 - T_s^4)(0.25)/[(1/e_1) + A_1/A_2((1/e_2) - 1)] \text{ W}$$

substituting data:

$$Q = \sigma(473^4 - T_s^4)(0.25)/[(1/0.95) + 0.25/0.5((1/0.2) - 1]$$
$$= \sigma(473^4 - T_s^4)(0.25)/3.0526$$

A heat balance can now be drawn up if temperatures remain steady:

Heat transfer from the pipe to the casing
 = heat transfer from the casing to the room

Thus

$$\sigma(473^4 - T_s^4)(0.25)/3.0526 = 0.2\sigma(T_s^4 - 293^4)(0.5)$$
$$473^4 - T_s^4 = 1.221(T_s^4 - 293^4)$$
$$5.9054 \times 10^{10} = 2.221 T_s^4$$

from which $T_s = 404$ K $\quad (t_s = 131°C)$

Thus the radiant heat transfer from the aluminium casing to the room will be:

$$Q = 0.2 \times 5.67 \times 10^{-8}(404^4 - 293^4)(0.5) = 109 \text{ W}$$

94 Heat radiation

SUMMARY FOR EXAMPLE 4.4

Condition of pipe	Radiant heat transfer (W)
Plain	575
Painted with aluminium	425
Wrapped in foil	120
Encased with aluminium cylinder	109

Example 4.5 Determination of the efficiency of a gas fired radiant heater

A gas fired radiant heater consumes 5.625 m³ in one and a half hours of natural gas which has a calorific value of 38.4 MJ/m³. The heater has an effective black body temperature of 750°C in surroundings at 20°C. If the area of the heater surface is 0.5 m² determine the radiant efficiency of the heater.

Solution
From equation 4.2 radiant heat flux $I = \sigma T^4$ W/m². For black body radiation exchange assuming the enclosure is black:

$$I = \sigma(T_1^4 - T_2^4) \text{ W/m}^2 \tag{4.8}$$

from which

$$Q = \sigma(T_1^4 - T_2^4)A_1 \text{ W}$$

substituting:

$$Q = 5.67 \times 10^{-8}(1023^4 - 293^4) \times 0.5$$

from which radiant output

$$Q = 31 \text{ kW}$$

Gas input $= [5.625/(3600 \times 1.5)] \times 38\,400 = 40\,\text{kW}$

Radiant efficiency $=$ (output/input) $\times$ 100
$\qquad\qquad\qquad\quad = (31/40) \times 100 = 77.5\%$

Example 4.6 Determination of the convection coefficient between flue gas and stem thermometer

A stem thermometer 4 mm in diameter is located in a bend along the axis of a boiler smoke pipe 200 mm in diameter. Flue gas temperature is 200°C and the reading on the thermometer is 185°C. The surface temperature of the smoke pipe is 140°C. Determine the convection coefficient for heat transfer between the flue gas and the stem thermometer. Emissivity of the stem thermometer is 0.93 and that of the smoke pipe 0.8.

Solution
There is a temperature disparity here between the flue gas, the stem thermometer and the wall of the boiler smokepipe and the reason for it is presented in the summary at the conclusion of the solution. A heat balance may be drawn up as follows:

Convection to the thermometer from the flue gas
= the radiation exchange between the thermometer and
the wall of the smoke pipe

Absolute temperatures are used and the flue gas is 473 K, the stem thermometer is 458 K and the smoke pipe wall is 413 K. Considering the heat radiation exchange, the form factor for concentric cylinders is found in equation 4.10

$$1/F_{1,2} = (1/0.93) + \{(\pi \cdot 0.004L)/[(\pi \cdot 0.2L)((1/0.8) - 1)]\}$$

from which $1/F_{1,2} = 1.0803$ and therefore $F_{1,2} = 0.926$.
Adapting equation 4.11

$$Q = (F_{1,2})\sigma(T_1^4 - T_2^4)A_1$$

Substituting the values

$$Q = 0.926 \times 5.67 \times 10^{-8}(458^4 - 413^4)A_1 \text{ W}$$

Considering the heat transfer by convection, from equation 3.16, $Q = h_c \cdot A \cdot dt$ W and substituting values $Q = h_c \cdot A_1 \cdot (473 - 458)$ W.
Thus combining the formulae into the heat balance

$$h_c \cdot A_1 \cdot (473 - 458) = 0.926 \times 5.67 \times 10^{-8}(458^4 - 413^4)A_1$$

from which $15h_c = 782.69$ and $h_c = 52.18$ W/m² K.

SUMMARY FOR EXAMPLE 4.6
The reason why the stem thermometer, flue gas and smoke pipe wall were not at the same temperature was because the boiler smoke pipe at 140°C which is 60 K below the flue gas temperature must be rapidly loosing heat to its surroundings. The thermometer would register a truer reading if the smoke pipe was adequately insulated.

96 Heat radiation

A well-insulated boiler flue pipe will minimise heat transfer between the flue gas, thermometer and smoke pipe wall. This analysis applies to other similar applications and accounts for errors in temperature measurement.

4.12 Asymmetric heat radiation

There are three cases of asymmetric heat radiation:

- Local cooling – radiation exchange with an adjacent cold surface as with a single glazed window.
- Local heating – radiation exchange with an adjacent hot surface or a series of point sources with spot lamps.
- Intrusion of short wave radiation as with solar radiation through glazing.

Unless all the inner surfaces of an enclosure are at the same temperature mean radiant temperature t_r will vary throughout the space This variation will produce a change in comfort temperature t_c and introduce asymmetry. Strong asymmetry will promote discomfort.

To quantify the degree of discomfort it is helpful to introduce two concepts:

- Plane radiant temperature (prt)
- Vector radiant temperature (vrt)

Plane radiant temperature is associated with the effects of radiant cooling which can result, for example, in front of an external single glazed window. Discomfort may result if the prt when facing the cold window surface is 8 K below the room comfort temperature t_c. Refer to Figure 4.11. An example of heating discomfort may occur if the vrt resulting from solar irradiation through a glazed window is greater than 10 K above the room comfort temperature.

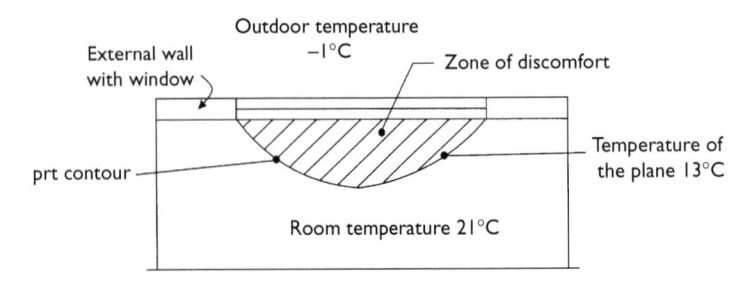

Figure 4.11 Plan of room with an external window showing prt contour.

In buildings which are highly intermittent in use such as churches and in factories where the air temperatures are low, directional high temperature radiant heaters giving a vrt well in excess of 10 K are quite acceptable since they compensate for the low air temperatures and low mean radiant temperatures resulting from the cold enclosing surfaces.

4.13 Historical references

L. Boltzman, Wiedemanns Annalen 22, 291, 1884.
H.C. Hottel, Notes on Radiant Heat Transmission Chemical Engineering Department, MIT, 1951.
G. Kirchoff, Ostwalds Klassiker d. exakten Wissens., 100, Leipzig, 1898.
H.L. Lambert, *Photometria*, 1860.
M. Planck, *The Theory of Heat Radiation* (translation), Dover, 1959.
S.J. Sitzungsber, *Akad. Wiss. Wien. Math - naturw.* Kl., 79, 391, 1879.

Note: Stefan established the constant σ experimentally. Boltzman subsequently proved it theoretically.

4.14 Chapter closure

You now have an underpinning knowledge of this mode of heat transfer and have investigated the use of luminous and non-luminous radiant heaters.

You understand the importance of the surface characteristics of the radiator and of its location in space for effective results. The concept of asymmetric radiation and discomfort has also been addressed.

Some work has been done in this chapter on saving energy by the use of bright aluminium foil. This is an important issue in relation to the reduction of carbon dioxide emission into the atmosphere.

Heat radiation at the wavelength of maximum flux has been shown to represent a significant proportion of the total.

Chapter 5

Measurement of fluid flow

Nomenclature

a	cross-sectional area (m^2)
C	constant
C_d	coefficient of discharge
dh	difference in head (m)
dP	pressure difference (Pa)
g	gravitational acceleration taken as 9.81 m/s^2 at sea level
h	head, metres of fluid flowing (m)
L	length of inclined scale (mm)
m	ratio of cross-sectional areas
M	mass transfer (kg/s)
P	pressure (Pa)
P_s	static pressure (Pa)
P_t	total pressure (Pa)
P_u	velocity pressure (Pa)
Q	volume flow rate (m^3/s)
R	radius of a circle (m)
S	ratio of densities
T	absolute temperature (K)
u	mean velocity (m/s)
x, H	vertical height (m)
Z	height above a datum (m)
ρ	density (kg/m^3)

5.1 Introduction

This chapter focuses upon the traditional instruments used for measuring gauge pressure, differential pressure and volume flow rate. It also considers the calibration of pressure measuring instruments.

5.2 Flow characteristics

It is helpful to begin with some general definitions relating to the flow of fluids in pipes and ducts.

Uniform flow. The area of cross section and the mean velocity of the fluid in motion are the same at each successive cross section, for example, flow of water through a flooded pipe of uniform bore.

Volume flow rate $Q = a \times u$ m^3/s

Steady flow. The area of cross section and the mean velocity of the fluid may vary from one cross section to the next but for each cross section they do not change with time, for example, flow of water through a flooded tapering pipe.

Volume flow rate $Q = a_1 \cdot u_1 = a_2 \cdot u_2$ m^3/s

Thus $u_2 = u_1 \cdot (a_1/a_2)$ m/s

Continuity of flow. The total amount of fluid entering and leaving a system of pipework or ductwork is the same. This occurs in uniform flow and steady flow, for example, air flow through a tee piece or junction. See Figure 5.1 in which $Q_1 = Q_2 + Q_3$. Furthermore, $u_1 \cdot a_1 = (u_2 \cdot a_2) + (u_3 \cdot a_3)$

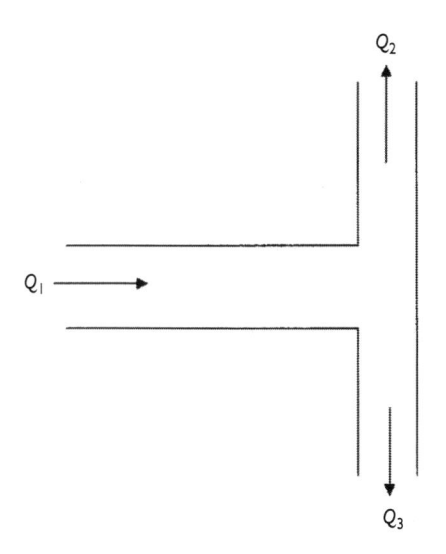

Figure 5.1 Continuity of flow $Q_1 = Q_2 + Q_3$.

100 Measurement of fluid flow

Mean velocity. Mean velocity u at any cross section of area a when the volume flow rate in m^3/s is Q will be $u = Q/a$ m/s.

5.3 Conservation of energy in a moving fluid

In order to consider the traditional methods of fluid flow measurement, it is necessary to introduce the Bernoulli equation which states that for frictionless flow:

potential energy Z + pressure energy $P/\rho g$ + kinetic energy $u^2/2g$
= a constant

In this format each energy term in the equation has the units of metres of fluid flowing. Thus for frictionless flow the total energy of the fluid flowing remains constant; no energy is lost or gained in the process.

- Potential energy is that due to a height above a datum, for example, water stored in a water tower has potential energy when ground level is taken as datum.
- Pressure energy is that due to static pressure or pump or fan pressure when present, for example, water flowing in a heating system subject to the sum of the static head imposed by the feed and expansion tank and also to the pressure developed by the pump.
- Kinetic energy is that due to the velocity of the fluid in the pipe or duct.

If two points are considered in a system in which fluid is flowing, one downstream of the other as shown in Figure 5.2, the following statement, assuming frictionless flow, can be made:

$$Z_1 + (P_1/\rho g) + (u_1^2/2g) = Z_2 + (P_2/\rho g) + (u_2^2/2g)$$

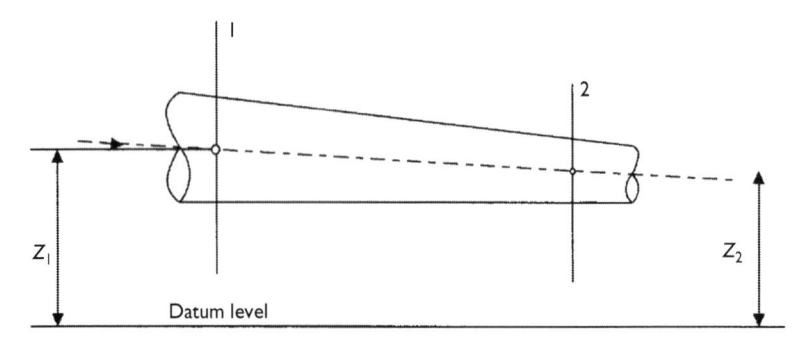

Figure 5.2 Conservation of energy in frictionless flow.

5.4 Measurement of gauge pressure with an uncalibrated manometer

Consider Figure 5.3 which shows a manometer open to atmosphere connected to a circular duct. At section A–A, the pressures are equal in each limb of the manometer and therefore:

pressure in the left-hand limb = pressure in the right-hand limb

Thus

$$(h + H)\rho_1 \cdot g = x \cdot \rho_2 \cdot g$$

gravitational acceleration g cancels and

$$(h + H) = x \cdot \rho_2/\rho_1$$

If $S = \rho_2/\rho_1$

$$h = S \cdot x - H \text{ m of fluid flowing}$$

The pressure of the fluid flowing

$$P = (S \cdot x - H)\rho_1 \cdot g \text{ Pa} \tag{5.1}$$

Example 5.1 Determining static pressure of air using a manometer

(a) A water filled manometer is connected to a duct through which air is flowing. If the displacement of water level is 43 mm, determine the static pressure generated by the air.
  Data: Water density 1000 kg/m^3, air density 1.2 kg/m^3 and $H = 350$ mm.
(b) Calibrate the manometer.

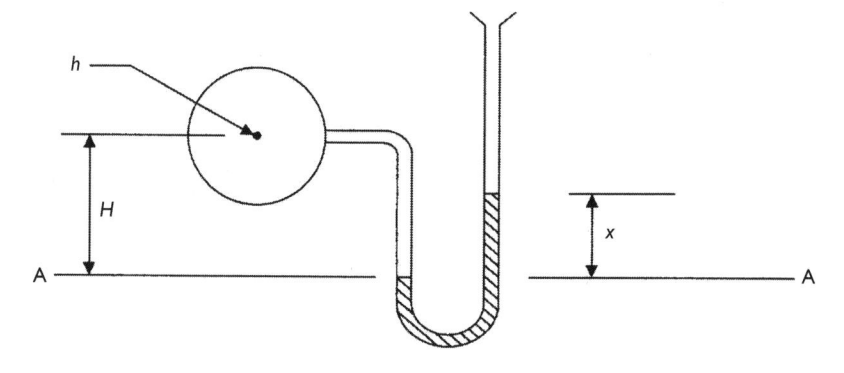

Figure 5.3 The manometer.

Solution (a)

The ratio of densities $S = 1000/1.2 = 833$. Substituting into equation 5.1

$$P = [(833 \times 0.043) - 0.35] \times 1.2 \times 9.81 = 418 \text{ Pa}$$

Note: Since the fluid flowing is air which has a relatively low density, H can be ignored without loss of integrity and $P = S \cdot x \cdot \rho_1 \cdot g$ Pa.

Solution (b)

The displacement of water of 43 mm is equivalent to a static pressure of 418 Pa. Thus, $418/43 = 9.72$ Pa per mm of water displacement. This is a displacement of approximately 1 mm of water for 10 Pa of static pressure and on this basis, the manometer can now be calibrated.

5.5 Measurement of pressure difference with an uncalibrated differential manometer

Consider Figure 5.4 which shows a differential manometer connected to a pipe transporting fluid. At section A–A, the pressures in each limb of the manometer are equal and pressure in the left-hand limb = pressure in the right-hand limb. Thus,

$$(h_1 + H)\rho_1 \cdot g = (h_2 + H - x)\rho_1 \cdot g + x \cdot \rho_2 \cdot g$$

Gravitational acceleration g cancels as does H, then

$$h_1 \rho_1 = (h_2 - x)\rho_1 + x\rho_2$$
$$h_1 \rho_1 = h_2 \rho_1 - x\rho_1 + x\rho_2$$

and $(h_1 - h_2)\rho_1 = x\rho_2 - x\rho_1$

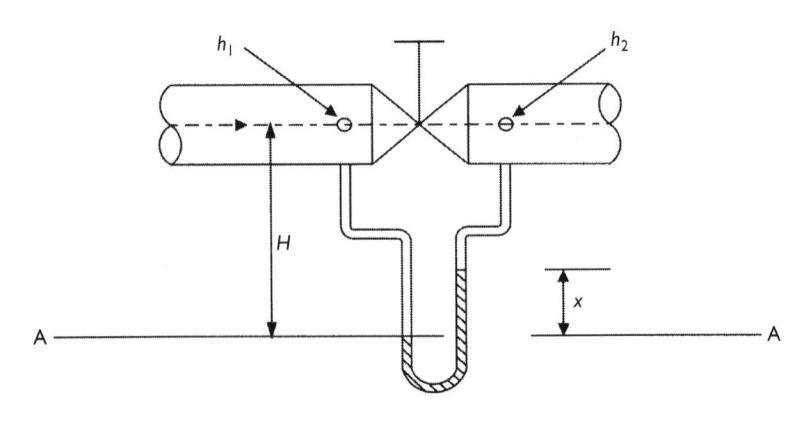

Figure 5.4 The differential manometer.

Therefore, $h_1 - h_2 = x(\rho_2/\rho_1) - x$

Then head loss $h_1 - h_2 = (x \cdot S - x) = x(S - 1)$ m of fluid flowing

or $dh = x(S - 1)$ m of fluid flowing (5.2)

Therefore pressure loss $dP = x(S - 1)\rho_1 \cdot g$ Pa (5.3)

Example 5.2 Determining whether an air filter needs replacing using a differential manometer

(a) A differential manometer measures the pressure drop across an air filter. The displacement of measuring liquid in the instrument is found to be 10 mm. Determine whether the filter should be replaced.
    Data: Density of measuring liquid 850 kg/m³, air density 1.2 kg/m³, maximum pressure drop across the filter when it should be replaced 50 Pa.
(b) Calibrate the manometer.

Solution (a)
The ratio of densities $S = 850/1.2 = 708$. Substituting the data into equation 5.3

$dP = 0.010(708 - 1) \times 1.2 \times 9.81 = 83.23$ Pa

Note: Unless the ratio of densities S has a relatively low value, equation 5.3 can be reduced to $dP = x \cdot S \cdot \rho_1 \cdot g$ Pa without loss of integrity.

Clearly the filter is in need of replacement since an excess pressure drop across it implies that it is partially clogged.

Solution (b)
The displacement of liquid of 10 mm results in a pressure drop of 83.23 Pa. Thus $83.23/10 = 8.32$ Pa per mm of displacement. This is a displacement of approximately 1 mm for 8 Pa of differential pressure and on this basis the instrument can now be calibrated.

Inclined differential manometers

The displacement of the measuring fluid has been considered in a limb of the manometer which is in the vertical position, see Figure 5.4. The calibration can be difficult to read accurately because of the small scale. Inclined manometers are used to ensure more accurate readings.

If the angle of inclination of the measuring limb is changed from the vertical to 20° from the horizontal, Figure 5.5, then $\cosine 70 = 1/L$ from which $L = 1/\cosine$ and $70 = 1/0.342 = 2.92$. Thus the scale length is now 2.9 times that of the corresponding vertical scale. Alternatively, 1 mm displacement on the vertical scale equals 2.9 mm on the scale inclined at 20° to the horizontal.

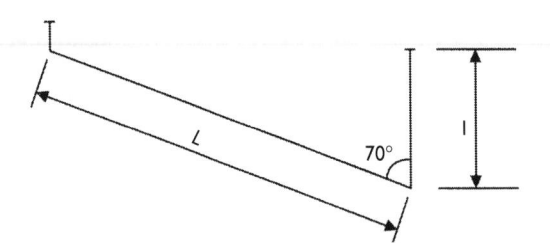

Figure 5.5 The inclined manometer, $L = 1 / \cos 70$.

In the case of the solution to Example 5.2, the displacement of water of 10 mm on the vertical scale of the differential manometer would now be extended on the inclined scale to $10 \times 2.9 = 29$ mm and the calibration would now be $83.23/29 = 2.87$ Pa per mm of displacement. You should now confirm that this is so.

5.6 Measurement of flow rate using a venturi meter and orifice plate

The venturi and orifice plate are instruments specifically made for each application. Once installed they are permanently fixed in position. Consider Figure 5.6 which shows a venturi fitted in a horizontal pipe in which fluid is flowing. The venturi is a fixed instrument and is purpose-made for the application.

The design of the venturi meter requires that the entry or converging cone has an angle of 21°, the length of the throat is equal to its diameter and the diverging cone has an angle of 5–7°. The two tappings measure static pressure and may be bosses or piezometer rings.

Applying the Bernoulli equation for frictionless flow at sections 1 and 2

$$Z_1 + (P_1/\rho \cdot g) + (u_1^2/2g) = Z_2 + (P_2/\rho \cdot g) + (u_2^2/2g)$$

Since the pipe is horizontal $Z_1 = Z_2$. Rearranging the equation,

$$(P_1 - P_2)/\rho \cdot g = (u_2^2 - u_1^2)/2g$$

We know the equation rearranges in this way since $u_2 > u_1$ thus $P_1 > P_2$. The equation may now be written as

$$dh = (u_2^2 - u_1^2)/2g$$

since $dh = dP/\rho \cdot g$.

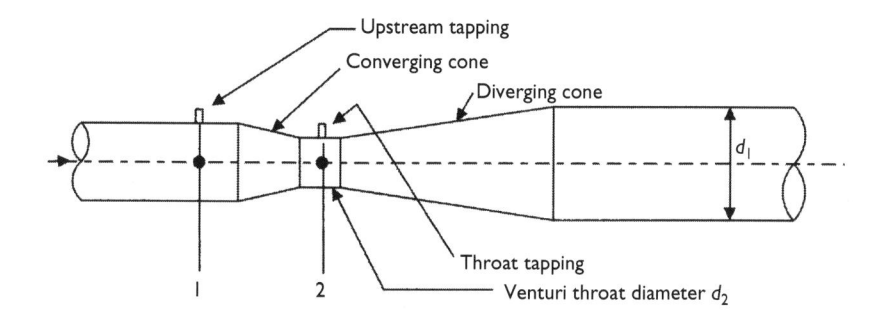

Figure 5.6 The venturi meter.

For continuity of flow, $a_1 \cdot u_1 = a_2 \cdot u_2$ in which d_1 and d_2 are fixed for the chosen application. So

$$u_2 = (a_1/a_2) \times u_1$$

If area ratio $m = a_1/a_2 = d_1^2/d_2^2$, $u_2 = m \cdot u_1$. Substituting

$$dh = [(m \cdot u_1)^2 - u_1^2]/2g$$

Then

$$dh = (u_1^2/2g)(m^2 - 1) \text{ m of fluid flowing}$$

In units of pressure

$$dP = (u_1^2/2g)(m^2 - 1)\rho \cdot g \text{ Pa}$$
$$= (\rho u_1^2/2)(m^2 - 1) \text{ Pa}$$

and rearranging

$$u_1 = [(2dP)/\rho(m^2 - 1)]^{0.5} \text{ m/s}$$

Therefore

$$Q_1 = u_1 \cdot a_1 = [(2dP)/\rho(m^2 - 1)]^{0.5} \times a_1$$

The formula is rearranged thus,

$$Q_1 = [2/\rho(m^2 - 1)]^{0.5} \times a_1 \times (dP)^{0.5}$$

106 Measurement of fluid flow

where $[2/\rho(m^2 - 1)]^{0.5} \times a_1 = C$, a constant for the instrument and based upon its physical dimensions for the chosen application and the density of the fluid flowing. Thus,

$$Q = C \cdot (dP)^{0.5} \text{ m}^3/\text{s}$$

This formula is derived from the Bernoulli equation for frictionless flow. Clearly, there will be a small loss due to friction as fluid passes through it. The coefficient of discharge C_d, determined for each instrument, accounts for this and for the venturi meter it varies between 0.96 and 0.98. It is found empirically before leaving the manufacturer. Therefore, actual flow

$$Q = C_d \cdot C \cdot (dP)^{0.5} \text{ m}^3/\text{s} \tag{5.4}$$

The orifice plate is shown in Figure 5.7. This also is a fixed instrument designed for a specific application. The determination of flow rate is obtained from the same equation as that for the venturi meter. The coefficient of discharge for the orifice plate C_d varies between 0.6 and 0.7.

Venturi meters are used for measuring the flow rate of liquids. The orifice plate is normally used for measuring the flow rate of gases, such as steam, since it has less effect upon the compressibility of the substance.

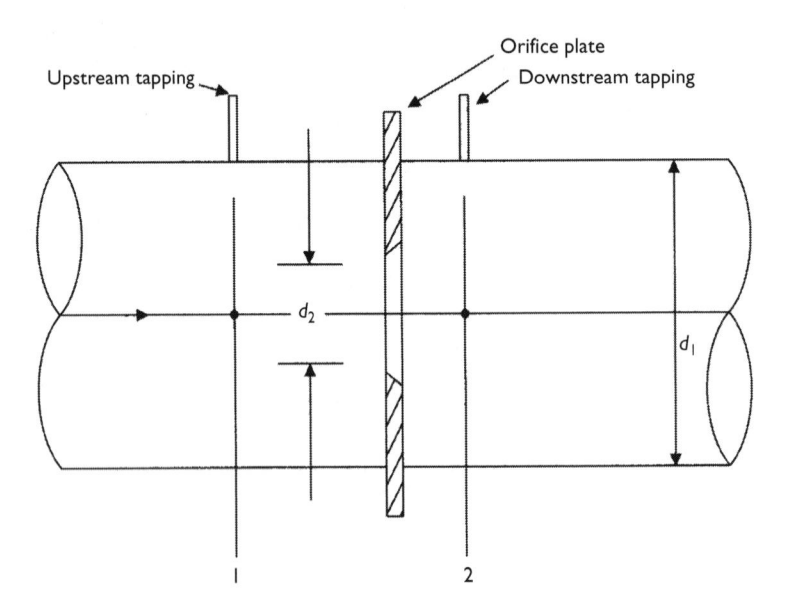

Figure 5.7 The orifice plate.

Example 5.3 Find the constant and the flow rate of water using an inclined differential manometer

A venturi meter is located in a horizontal pipeline transporting water at 75°C and is connected to an uncalibrated differential manometer whose calibration limb is inclined at 20° to the horizontal. From the data, find the constant C for the meter.

If the inclined manometer shows a displacement of measuring liquid of 70 mm, determine the volume flow and hence the mass transfer of water in the pipe.

Data: Pipe diameter 50 mm, venturi throat diameter 25 mm, water density 975 kg/m^3, density of measuring liquid is 13 600 kg/m^3 and the coefficient of discharge for the meter C_d is 0.96.

Solution
The manufacturer of the venturi would have requested details of the temperature of the water flowing, the design flow rate and the diameter of the pipe. The constant C would then have been supplied with the instrument. Here we are asked to calculate it. It was established that constant

$$C = [2/\rho(m^2 - 1)]^{0.5} \times a_1$$

where $m = d_1^2/d_2^2 = (50/25)^2 = 4$. Substituting:

$$C = [2/975(16 - 1)]^{0.5} \times (\pi \times 0.05^2/4) = 0.000023$$

Actual flow rate through the meter $Q = C_d \cdot C \cdot (dP)^{0.5}$ from equation 5.4. A differential manometer already calibrated in Pa or mbar would be used to measure the pressure drop across the tappings on the venturi meter. Here we have to calculate it.

The differential pressure loss dP for the water flowing is determined from equation 5.3. However, the vertical displacement x is required and since cosine $70 = x/L$, where L = the calibration limb at 20° to the horizontal, $x = L \cdot \cos 70$; from which $x = 70 \times 0.342 = 24$ mm. From equation 5.3, $dP = x(S - 1)\rho \cdot g$, where $S = 13\,600/975 = 13.95$. Thus

$$dP = 0.024(13.95 - 1) \times 975 \times 9.81 = 2973 \text{ Pa}$$

Substituting into equation 5.4

$$Q = 0.96 \times 0.000023 \times (2973)^{0.5} = 0.0012 \text{ m}^3/\text{s}$$
$$= 1.2 \text{ L/s}$$

and mass transfer $M = Q \cdot \rho = 0.0012 \times 975 = 1.174$ kg/s

108 Measurement of fluid flow

SUMMARY FOR EXAMPLE 5.3
The calibration for this differential manometer measuring water flowing at 75°C would be $2973/70 = 42.5$ Pa for each millimetre displacement of measuring fluid. Notice the effect of temperature on the density of the fluid flowing and therefore the differential pressure reading. If the water flowing was at 5°C its density would be 1000 kg/m^3 and

pressure loss $dP = 0.024(13.6 - 1) \times 1000 \times 9.81 = 2967$ Pa;
constant $C = [2/1000(16 - 1)]^{0.5} \times (\pi \times 0.05^2/4) = 0.0000227$;
flow rate $Q = 0.96 \times 0.0000227(2967)^{0.5} = 0.001187$ m^3/s; and finally
mass transfer $M = Q \cdot \rho = 0.001187 \times 1000 = 1.187$ kg/s.

Example 5.4 Calculate the rate of steam flow using an orifice plate

An orifice plate is installed in a steam main for measuring the flow of steam. Determine the rate of flow given the manufacturer's coefficient of discharge as 0.7 and the manufacturer's constant for the instrument as $C = 0.0046$. The differential pressure measured at the orifice plate was 290 mbar. Given the steam density as 3.666 kg/m^3, find the mass flow rate of steam in the pipe.

Solution
The manufacturer's constant C derives from $C = [2/\rho(m^2 - 1)]^{0.5} \times a_1$. The manufacturer must therefore be provided with the steam pressure and quality in order to establish its density, and the pipe diameter into which the orifice plate is to be fitted as a permanent device.

The measured pressure drop of 290 mbar = 29 000 Pa. From equation 5.4,

$$Q = C_d \cdot C \cdot (dP)^{0.5} \text{ m}^3/\text{s}$$

Substituting, the volume flow rate of steam

$$Q = 0.7 \times 0.0046 \times (29\,000)^{0.5} = 0.548 \text{ m}^3/\text{s}$$

The mass transfer of steam

$$M = 0.548 \times 3.666 = 2 \text{ kg/s}$$

Example 5.5 Sizing a recording chart for a venturi meter fitted to a water main

A venturi meter is fitted into a horizontal water main and is intended to act as a means of recording the water flow rate to a process. To do this, the pressure tappings are connected to the ends of a cylinder of 20 mm bore fitted with a piston which has a pen connected to the piston rod by means of a linkage in such a way that each millimetre of rod movement causes the pen to move 10 mm across the paper.

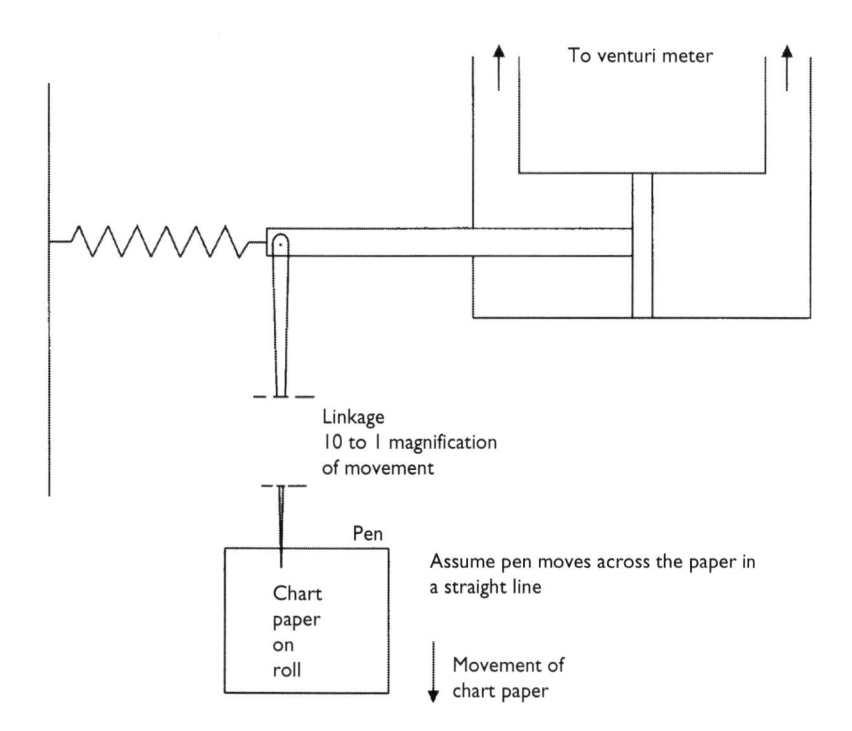

Figure 5.8 Use of venturi meter and chart recorder (Example 5.5).

The system is shown diagrammatically in Figure 5.8. The rate of water flow may vary between 240 and 170 L/s during the process operation. The water main is 300 mm bore and the throat of the venturi is 200 mm bore.

(a) Determine the velocity of water through the pipe and the force on the piston at each of the two flow rates. Ignore the diameter of the piston rod. Take water density as 1000 kg/m^3.
(b) If the spring extends 4 mm per Newton, determine the minimum width of chart paper needed to record the flow rates between the two limits.

Solution (a)
Since $Q = u \cdot a$, $u = Q/a = 4Q/\pi d^2$ and for a flow rate of 0.24 m^3/s

$$u_1 = 0.24 \times 4/\pi \times (0.3)^2 = 3.395 \text{ m/s} \quad \text{and}$$
$$u_2 = 0.24 \times 4/\pi \times (0.2)^2 = 7.639 \text{ m/s}$$

Similarly for a flow rate of 0.17 m^3/s, $u_1 = 2.405$ m/s and $u_2 = 5.411$ m/s.

110 Measurement of fluid flow

Adopting the Bernoulli equation in Section 5.3 and taking section 1 at the tapping on the upstream pipe and section 2 at the tapping on the throat of the venturi.

$$Z_1 + (P_1/\rho g) + (u_1^2/2g) = Z_2 + (P_2/\rho g) + (u_2^2/2g)$$

Since the venturi is horizontal $Z_1 = Z_2$. Since $d_1 > d_2$, $u_2 > u_1$ and rearranging the equation:

$$(P_1 - P_2)/\rho g = (u_2^2 - u_1^2)/2g$$

For a flow rate of 0.24 m³/s

$$(P_1 - P_2)/\rho g = [(7.639)^2 - (3.395)^2]/2g \quad \text{and} \quad dP = 23\,417 \text{ Pa}$$

Now dP = force/area. So force = dP × area of piston = $23\,417 \times \pi(0.02)^2/4 = 7.357$ N. Similarly for a flow rate of 0.17 m³/s, dP is calculated as 11 749 Pa and force = $11\,749 \times \pi(0.02)^2/4 = 3.69$ N. You should now confirm this calculation.

Solution (b)
Spring movement = $(7.357 - 3.69) \times 4 = 14.7$ mm. Linkage = $14.7 \times 10 = 147$ mm. Therefore chart width should be a minimum of 150 mm.

The inclined venturi

Normally the venturi meter is positioned horizontally. It is, however, pertinent to consider the effect upon a reading when the instrument is located in an inclined position. Figure 5.9 shows such a location. At section A–A, the head exerted in the left-hand limb = the head exerted in the right-hand limb. Thus,

$$P_1/\rho g + Z_1 = P_2/\rho g + (Z_2 - x) + Sx$$

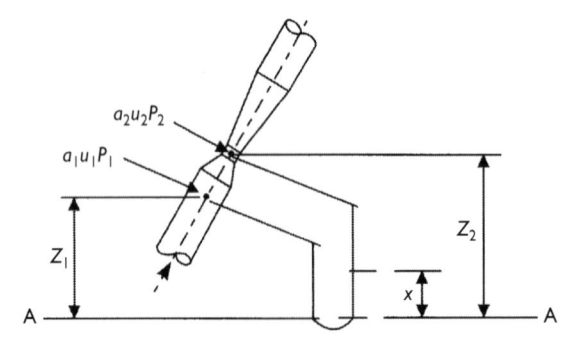

Figure 5.9 The inclined venturi meter.

Measurement of fluid flow

where S is the ratio of densities of the measuring fluid and the fluid flowing. Rearranging

$$[(P1 - P2)/\rho g] + Z_1 - Z_2 = Sx - x = x(S - 1) = dh$$

This is equation 5.2 for head loss in a differential manometer.

Thus, the final equation 5.4 is independent of Z_1 and Z_2 and there is no effect on the measurements taken from a venturi meter located in the inclined position.

5.7 Measurement of air flow using a pitot static tube

The pitot static tube unlike the venturi and orifice plate is a portable instrument and used to measure air flow. It consists of two tubes in a coaxial arrangement as shown in Figure 5.10. The inner tube connected to the nose of the instrument measures total pressure in the air stream. The outer tube has holes in its sides which measure the static pressure of the air stream. By connecting each tube to a differential manometer the velocity pressure of the air stream is obtained.

Consider the Bernoulli theorem for the total energy at a point in a system of air flow

$$(Z + P/\rho g + u^2/2g) \text{ in metres of air flowing} = \text{total energy in the air}$$

Z, the potential energy due to the height above a datum is insignificant because air density is relatively very low. If energy is measured in units of

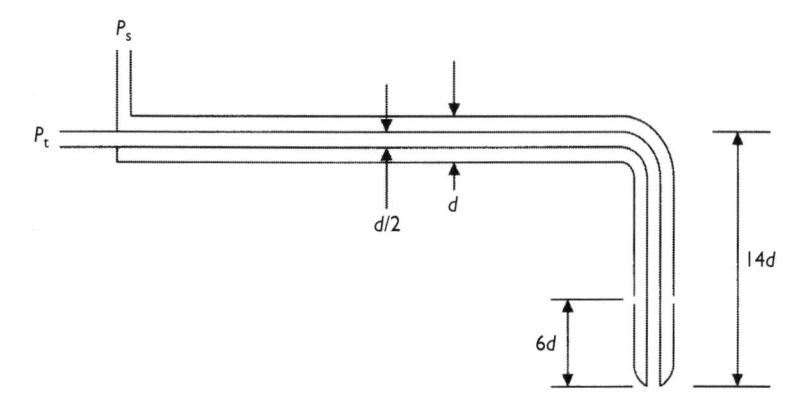

Figure 5.10 The pitot static tube.

pressure:

$$(P/\rho g)\rho g + (u^2/2g)\rho g = \text{total energy of the moving air in Pa}$$

Thus, $P + (\rho/2)u^2 = \text{total energy}$

The first term, P is the static pressure energy generated by the fan working on the air and the second term is the pressure energy due to the velocity of the moving air. These terms are commonly called static pressure P_s and velocity pressure P_u.

Thus total pressure of the moving air at a point $P_t = P_s + P_u$ Pa, where $P_u = (\rho/2)u^2$ Pa.

When the pitot tube is connected to a differential manometer velocity pressure is obtained and $P_u = P_t - P_s$. Figure 5.11 shows the static and total pressure tubes of the pitot tube separated to identify the equivalent readings.

Rearranging the formula for velocity pressure P_u, mean air velocity

$$u = (2P_u/\rho)^{0.5} \text{ m/s} \tag{5.5}$$

This is the theoretical velocity. Actual velocity $= C \cdot (2P_u/\rho)^{0.5}$ m/s and actual flow rate $Q = C \cdot (2P_u/\rho)^{0.5} \times a$ m^3/s.

For the pitot static tube constant $C = 1.0$ for Reynolds numbers greater than 3000 and where the cross-sectional area of the pitot tube is insignificant compared with the cross-sectional area of the duct a. Thus when air flow is in the turbulent region

$$Q = (2P_u/\rho)^{0.5} \times a \text{ m}^3/\text{s} \tag{5.6}$$

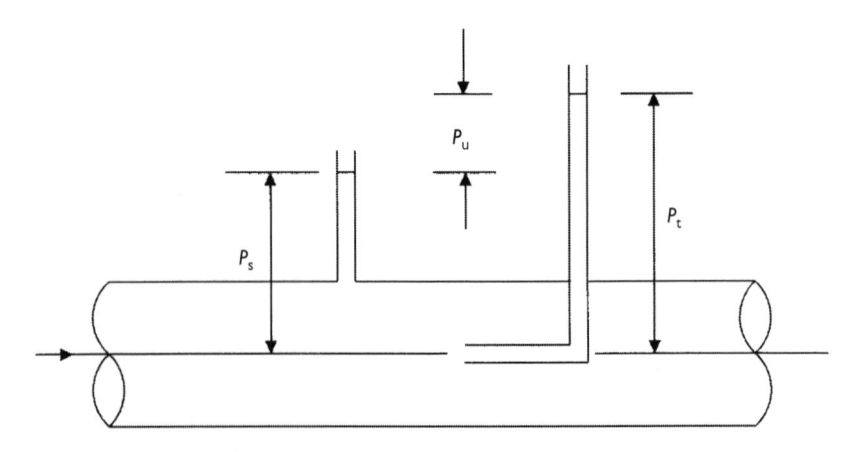

Figure 5.11 The equivalent readings from a pitot static tube.

Measurement of fluid flow

The accuracy of readings of velocity pressure depends very much on the person using the pitot tube ensuring that the nose of the instrument is pointing at the air stream and parallel to the duct. Readings are best taken on a straight section of duct where the velocity profile of the air is more likely to be regular.

Air velocity distribution

For square and rectangular ducts the cross section is divided into a grid so that the velocity pressure, velocity and hence volume flow Q is the sum of the flow rates in each segment of the grid, thus:

$$Q = a_1 \cdot u_1 + a_2 \cdot u_2 + a_3 \cdot u_3 + \cdots = \sum (a \cdot u) \text{ m}^3/\text{s}$$

and

$$\text{mean velocity} = \sum (a \cdot u) / \sum a \text{ m/s}$$

For circular ducts the velocity profile on straight sections is the same across any diameter, therefore, for an annular ring of mean radius r and area a, volume flow $Q = \sum (u_r \cdot a_r) \text{ m}^3/\text{s}$ and mean velocity $= \sum (u_r \cdot a_r)/\pi R^2$ m/s.

Air density varies with temperature and pressure. Standard air density ρ_1 at 20°C and 101 325 Pa is 1.2 kg/m^3 and for other temperatures and pressures air density ρ_2 can be found by applying the gas laws. Thus

$$\rho_2 = \rho_1 (T_1/T_2)(P_2/P_1) \text{ kg/m}^3 \qquad (5.7)$$

Air density can also be found from the tables of *Thermodynamic and Transport Properties of Fluids* at constant standard air pressure and different absolute temperatures.

Example 5.6   Determine the volume flow rate of air in a duct using a pitot static tube

Air is conveyed in a duct of section 300 × 450 mm. A series of readings of velocity pressure are obtained with the aid of a pitot static tube and calibrated manometer. The mean value is determined as 65 Pa. Determine the mean air velocity in the duct and hence the volume flow rate.

Data: Temperature of the air flowing 35°C, barometric pressure 748 mm mercury, density of mercury 13 600 kg/m^3.

Solution

Local atmospheric pressure

$$P = h \cdot \rho \cdot g = 0.748 \times 13\,600 \times 9.81$$
$$= 99\,795 \text{ Pa}$$

Absolute temperatures

$$T_1 = 273 + 20 = 293\,\text{K}, \quad T_2 = 273 + 35 = 308\,\text{K}$$

Atmospheric pressures

$$P_2 = 99\,795 \text{ Pa}, \quad P_1 = 101\,325 \text{ Pa}$$

From equation 5.7

$$\rho_2 = 1.2(293/308)(99\,795/101\,325)$$
$$= 1.124 \text{ kg/m}^3$$

From equation 5.5,

$$u = (2P_u/\rho)^{0.5} = (2 \times 65/1.124)^{0.5}$$
$$= 10.75 \text{ m/s}$$

From equation 5.6,

$$Q = 10.75 \times (0.3 \times 0.45) = 1.45 \text{ m}^3/\text{s}$$

5.8 Chapter closure

You are now able to determine gauge pressure, differential pressure and flow rate using the traditional pressure measuring instruments and flow measuring devices. The working limits of the pitot static tube have been discussed. You can also calibrate the pressure measuring instruments described in this chapter and the inclined manometer has been introduced to obtain more accurate readings. This work can now be extended to other examples and problems in the field of measurement of pressure, differential pressure and flow rate.

Chapter 6

Characteristics of laminar and turbulent flow

Nomenclature

A	area of cross section (m^2)
d	diameter (m, mm)
dP	pressure drop (Pa)
dp	specific pressure drop (Pa/m)
dx	unit length (m)
f	frictional coefficient in turbulent flow
g	gravitational acceleration (m/s^2)
k_s	absolute roughness (mm)
L	length (m)
M	mass transfer (kg/s)
P_w	power (W)
Q	volume flow rate (m^3/s)
R, r	radius (m)
x, L, d	characteristic dimension (m)
μ	absolute viscosity (kg/m·s)
ν	kinematic viscosity (m^2/s) = μ/ρ
ρ	density (kg/m^3)

6.1 Introduction

Fluid viscosity is the measure of the internal resistance sustained in a fluid being transported in a pipe or duct as one layer moves in relation to adjacent layers. At ambient temperature, heavy fuel oils, for example, possess a high viscosity while the lighter oils possess a low viscosity. The walls of the pipe or duct provide the solid boundaries for the fluid flowing and because of the friction generated between the boundary and the fluid interface, which has a drag effect, and fluid viscosity, the velocity of flow varies across the enclosing boundaries to produce a velocity gradient. In a straight pipe or duct, maximum fluid velocity would be expected to occur along the centreline and zero velocity at the boundary surfaces.

116 Laminar and turbulent flow

There are, therefore, two factors to consider when water, for example, flows along a pipe, namely the viscosity of the water and the coefficient of friction at the pipe inside surface. Fluid viscosity is temperature dependent and the coefficient of friction at the inside surface of the pipe or duct is velocity dependent as well as being related to surface roughness and a characteristic dimension of the pipe, namely pipe diameter. This sets the scene for a discussion on laminar and turbulent flow.

6.2 Laminar flow

In about 1840 a Frenchman by the name of Poiseuille and an American by the name of Hagen identified the following equation which is dedicated to them during experiments on fluid viscosity:

$$Q = (\pi r^4 dP)/(8\mu dx) \text{ m}^3/\text{s} \tag{6.1}$$

This formula can be rearranged in terms of pressure drop per metre run of pipe or duct dp. Thus:

$$(dP/dx) = dp = 8Q\mu/\pi r^4$$

Since $Q = u \cdot A$, where $A = \pi r^2$. Then by substitution $Q = u \cdot \pi r^2$.
Substituting for Q, $dp = 8\mu u/r^2$ and since $r = d/2$ then $dp = 8\mu u/(d/2)^2$. Thus,

$$dp = 32\mu u/d^2 \text{ Pa/m} \tag{6.2}$$

and $$dP = 32\mu ux/d^2 \text{ Pa} \tag{6.3}$$

This equation can be expressed in terms of head loss dh of fluid flowing and since $dP = dh \cdot \rho \cdot g$ Pa, and substituting this for dP

$$dh = 32\mu ux/\rho g d^2 \text{ m of fluid flowing} \tag{6.4}$$

This arrangement of the Poiseuille/Hagen formula is probably better known than equation 6.1. You will notice that the effects of friction between the fluid and the boundary interface is not accounted for. This is because the dominating feature resisting fluid flow is the viscosity of the fluid. Laminar flow is therefore sometimes referred to as viscous flow.

It is not too clear whether Hagen or Poiseuille fully understood the characteristics of laminar flow during the process of establishing their formula, which shows that head loss dh is proportional to the ratio of u/d^2. This is verified by their reaction to the claim by another Frenchman called D'Arcy who in about 1857 proposed a different formula in which head loss due to friction is proportional to the ratio of u^2/d. It was left to an Englishman by the name of Osborne Reynolds to reconcile the dispute in 1883 at his famous presentation to the Royal Society. Refer to Figure 6.1.

Reynolds established that D'Arcy on the one hand and Hagen/Poiseuille on the other were both correct and that the different formulae were the result of different types of fluid flow. He found that in laminar flow the fluid moved along streamlines which are parallel to the pipe wall. See Figures 6.2 and 6.3. Any disturbance in a straight pipe in which the fluid was moving in laminar flow would cause a disturbance along the streamline which would dissipate at some point downstream and return to a streamline.

The experiment which Reynolds presented to the meeting of the Royal Society used coloured dye to illustrate the phenomenon. He was able to show that as fluid velocity increased, the streamline could not be maintained

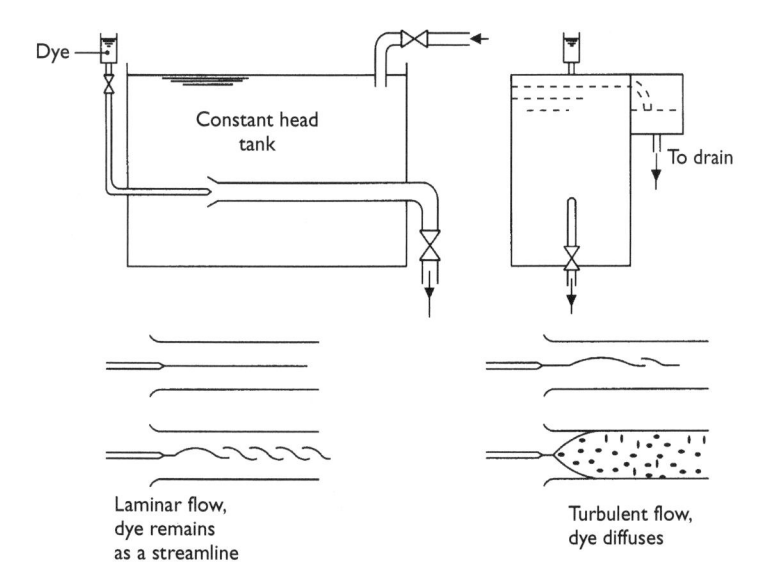

Figure 6.1 Osborne Reynolds' experiment (1883).

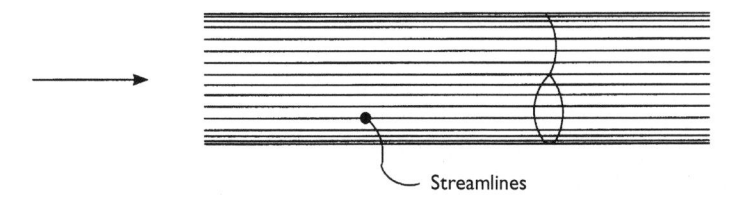

Figure 6.2 Streamlines in laminar flow.

118 Laminar and turbulent flow

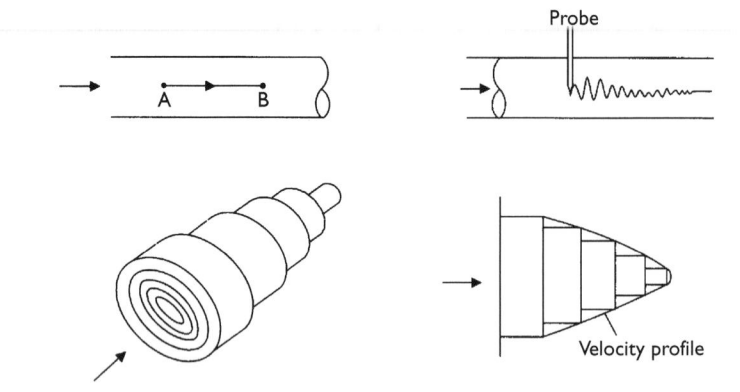

Figure 6.3 Laminar flow.

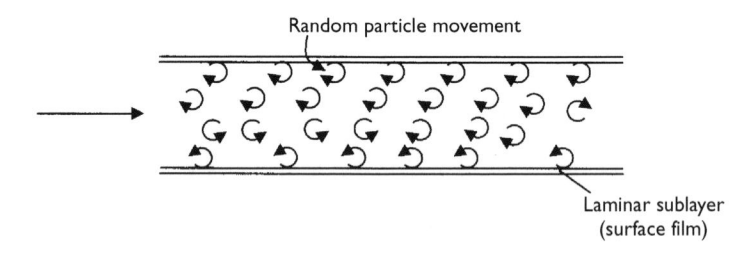

Figure 6.4 Fluid in turbulent flow.

and the point was reached when the dye suddenly diffused in the water which was used for the experiment and flow became turbulent.

Turbulent flow therefore can be identified as the random movement of fluid particles in a pipe or duct with the sum of the movements being in one direction. Refer to Figure 6.4.

Laminar flow characteristics

These are illustrated in Figures 6.1, 6.2, 6.3 and 6.5.

- A particle at point A at some time will be at point B after travelling in a straight line parallel to the tube.
- A disturbance in the fluid generated by the insertion of a probe will straighten out downstream.
- The change in velocity of the fluid across the pipe section is not linear but parabolic.

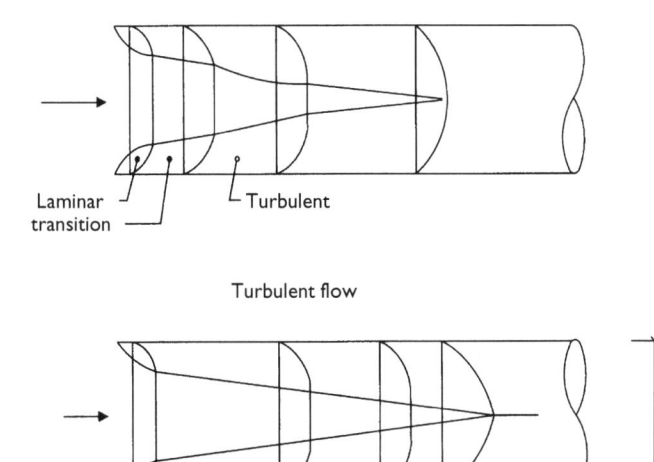

Figure 6.5 Development of velocity profiles – flow in pipes.

- It is convenient to think of the fluid motion as a series of concentric layers slipping over one another and the distances by which each layer is extruded represents the velocity of each layer.

6.3 Turbulent flow

As fluid velocity is increased in the pipe or duct the flow changes from laminar to turbulent. This is known as the critical point. If the velocity of fluid through a pipe in which flow is known to be laminar is increased slowly until flow just becomes turbulent the higher critical point is reached. If now the velocity is slowly decreased, flow will at first remain turbulent then at a velocity lower than that at which turbulence commenced, flow will again become laminar. This is known as the lower critical point.

Between the higher critical point and the lower critical point fluid flow is unstable. The Frenchman D'Arcy had identified a formula for turbulent flow in 1857 which, before the work done by Reynolds, could not be reconciled with the formula of Poiseuille and Hagen. The D'Arcy equation is:

$$dh = 4fLu^2/2gd \text{ m of fluid flowing} \tag{6.5}$$

Analysis of this formula is included in Appendix 2, Case study A2.1.

In the D'Arcy formula, fluid viscosity μ is not present but frictional coefficient f is. Here the dominant feature is the frictional resistance to flow at the boundary surface and the effects of the fluid viscosity in terms of resistance to flow are insignificant. Turbulent flow for this reason is sometimes called frictional flow.

Osborne Reynolds found that a dimensionless group of variables could be used to reconcile equations 6.4 and 6.5 and his name is used to identify the dimensionless group as the Reynolds number Re.

$$Re = \rho u d/\mu \tag{6.6}$$

It can also be expressed as

$$Re = dM/\mu A = ud/\nu \tag{6.7}$$

Reynolds' experiments identified the following general guidelines: for pipes and circular ducts when $Re < 2000$, flow is said to be in the laminar region and equation 6.4 can be adopted. Between an Re of 2000 and 3500, flow is in transition and therefore unstable. Above an Re of 10 000 flow is said to be turbulent and the D'Arcy equation 6.5 may be used. For practical calculations, equation 6.5 is also used for Reynolds numbers in excess of 3500.

The D'Arcy equation for turbulent flow can be expressed in terms of volume flow rate in a similar way to the Poiseuille/Hagen equation 6.1.

$$dh = 4fLu^2/2gd \text{ m of fluid flowing}$$

Since $Q = u \cdot A$, $u = Q/A = 4Q/\pi d^2$. Substituting for u in the D'Arcy equation:

$$dh = (4fL(4Q/\pi d^2)^2)/2gd = (64/2g\pi^2)(fLQ^2/d^5)$$

If the term $(64/2g\pi^2)$ is treated as constant it reduces to 0.33 so $dh = (1/3)fLQ^2/d^5$, from which

$$Q = (3dh \cdot d^5/fL)^{0.5} \text{ m}^3/\text{s} \tag{6.8}$$

This is known as Box's formula.

Turbulent flow characteristics

These are illustrated in Figures 6.1, 6.4 and 6.5.

- When the fluid velocity is high, disturbances in the fluid are not damped out.
- Fluid particles travelling along the pipe also travel across it in a random manner.

- Fluid particles cannot pass through the pipe wall and as the pipe surface is approached these perpendicular movements must die out. Thus turbulent flow cannot exist immediately in contact with the solid boundary.
- Even when the mean velocity is high resulting in a high Re number and the greater part of the boundary layer is turbulent, there remains a very thin layer adjacent to the solid boundary in which flow is laminar. This is called the laminar sublayer.

Summary

Most if not all fluid flow in the context of building services is in the turbulent region although it is well to check that this is so from the Reynolds number before proceeding with the solution to a problem.

6.4 Boundary layer theory

The velocity of a mass of fluid in motion which is subject to gravity and is remote from solid boundaries is uniform and streamline. There is no velocity gradient and hence there is no shear stress in the fluid. The viscosity of the fluid is therefore not affecting fluid motion and neither is the friction at the fluid boundary interface. Fluid in contact with a solid boundary is brought to rest. Further away, fluid will be slowed but by not as much as that closer to the boundary. Thus near solid boundary surfaces the effects of friction and fluid viscosity result in a velocity gradient. Refer to Figure 6.6.

For fluid flow in flooded pipes or ducts, the pipe/duct walls act as the solid boundary where fluid velocity is said to be zero as a result of fluid viscosity and the frictional resistance at the fluid/boundary interface.

L. Prandtl, who is considered the founder of fluid mechanics, defined the theory of the boundary layer in a variety of applications at the turn of the

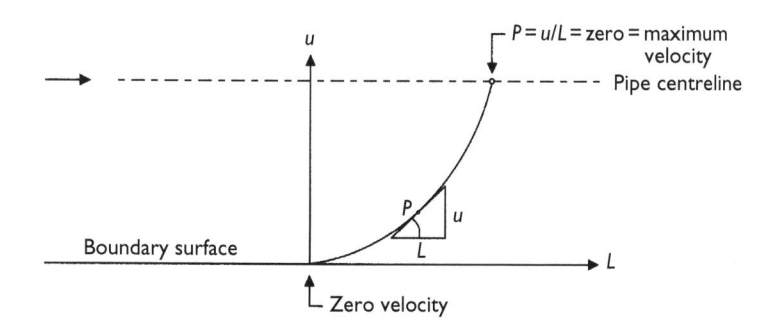

Figure 6.6 Velocity gradient – fluid flow in circular conduits.

122 Laminar and turbulent flow

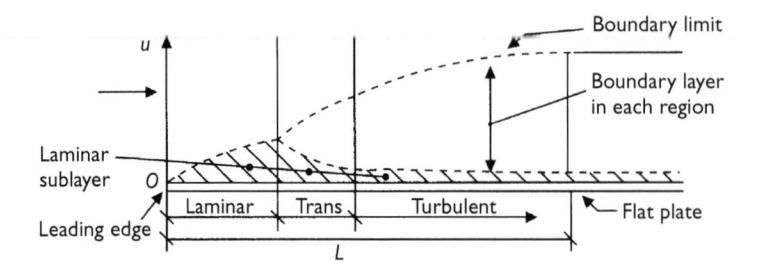

Figure 6.7 Formation of the boundary layer on one side of a flat plate.

century in Hanover and subsequently in Gottingen where he founded the Kaiser Wilhelm Institute. The building services engineer is mainly concerned with a limited number of applications such as the flow of air in ducts and the flow of water in pipes and channels. If a flat plate is positioned in a stream of flowing fluid which is unaffected by solid boundaries, the development of the boundary layer from the leading edge of the plate can be identified, one side being considered. Refer to Figure 6.7. The following points can then be observed:

- Fluid velocity under the boundary layer starts at zero at the leading edge of the plate and reaches a maximum at the boundary limit.
- The thickness of the boundary layer is very small compared with its length L.
- There are three discrete regions.
- Laminar and transition lengths are very short; flow therefore is often considered turbulent throughout the whole boundary layer.
- During transition Re has critical values.
- The plate imposes a resistance to flow causing a loss in fluid momentum. The plate experiences a corresponding force called skin friction.
- The boundary layer increases in thickness to a maximum value as the length L from the leading edge of the plate increases.
- At points close to the solid boundary of the flat plate, velocity gradients are large and the viscous shear mechanism is significant enough to transmit the shear stress to the boundary such that the layer adjacent to the boundary is in laminar motion even when the rest of the boundary layer is turbulent. This is the laminar sublayer which you will notice becomes extremely thin downstream of the leading edge of the flat plate.

- A pipe may be considered as a flat plate wrapped round to reform itself. Thus fluid velocity starts at zero at the pipe wall and reaches a maximum value at the centreline of a straight pipe. It then returns to zero velocity at the opposite wall of the pipe forming the velocity profile which is bullet shaped. The length L from the leading edge of the flat plate becomes infinite when the plate is reformed into a pipe since the thickness of the boundary layer is restricted at the pipe centreline by the boundary layer from the pipe wall opposite. If the pipe or duct was very large in diameter this may not be so. The length L, therefore, for most practical applications becomes the straight length of the pipe or duct being considered.

Figure 6.8 shows the formation of the boundary layer in laminar and turbulent flow in a straight pipe.

Velocity profile for laminar and turbulent flow in straight pipes

Due to the surface resistance at the boundary walls of the pipe and the viscosity of the fluid, maximum velocity occurs at the pipe centreline and zero velocity at the pipe wall. The velocity gradient u/L may be obtained at any point P on the velocity profile, Figure 6.6.

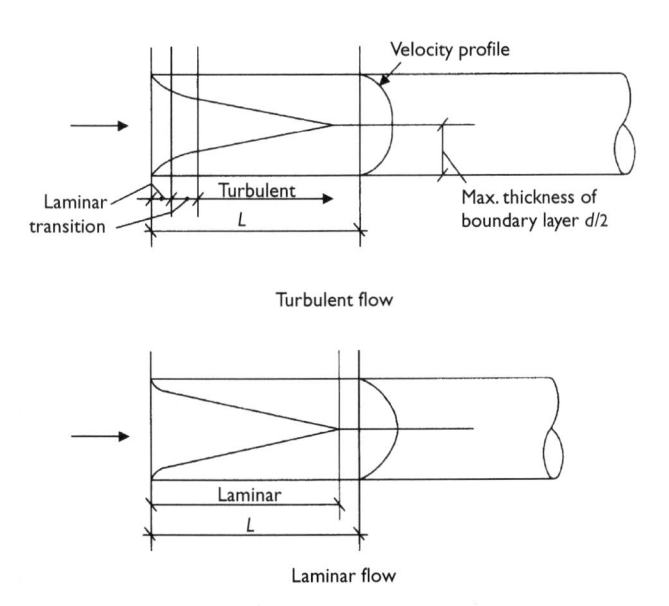

Figure 6.8 Formation of the boundary layer in a pipe.

124 Laminar and turbulent flow

At the pipe centreline the velocity profile $u/L =$ zero. Thus the boundary layer is the layer of fluid contained in a velocity profile up to the point where the velocity gradient is zero.

Boundary layer separation

The separation of the boundary layer from the solid boundary surface does not occur in straight pipes or ducts. This is because there is a steady static pressure loss in the direction of flow. It does occur, however, in tees, Y junctions, bends and gradual enlargements, and its effects on pressure losses through fittings are analysed in Chapter 7.

It can be shown that in each of the fittings identified here, there is a momentary gain in static pressure as the fluid passes through. This is most commonly noted in the gradual enlargement in which the gain in static pressure is held. The gain in static pressure is at the expense of a corresponding loss in velocity pressure whether it is momentary or otherwise and this causes the boundary layer to separate from the solid boundary surface. It rejoins at some point downstream. Figures 6.9–6.12 illustrate the phenomenon.

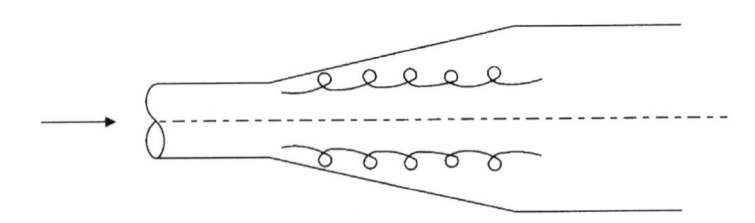

Figure 6.9 Boundary separation in an enlargement.

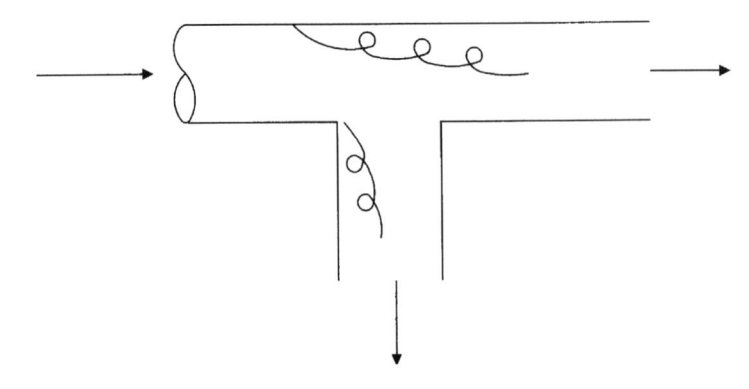

Figure 6.10 Boundary separation in a tee piece.

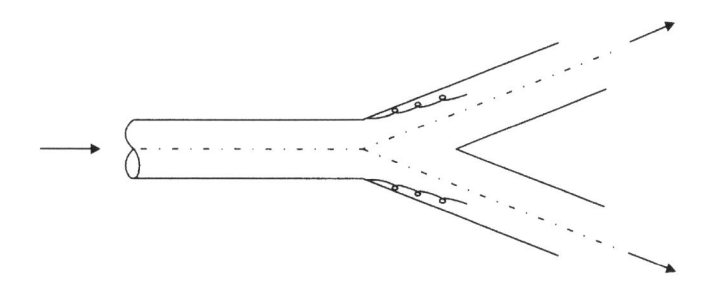

Figure 6.11 Boundary separation in a Y junction.

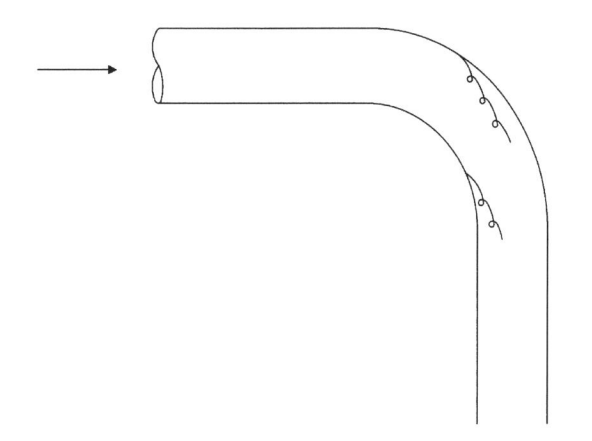

Figure 6.12 Boundary separation in a 90° bend.

6.5 Characteristics of the straight pipe or duct

The coefficient of friction f appears in the D'Arcy equation 6.5 but does not figure in equation 6.4 of Poiseuille/Hagen. The reason for the omission is because the roughness of the pipe wall is not a significant factor in laminar flow. The coefficient of friction f at the fluid boundary is a function of a lineal measurement of the high points on the rough internal surface k_s called surface roughness/absolute roughness and measured in mm. It is also a function of a characteristic dimension of the pipe, taken as its diameter d, or in the case of a rectangular section the shorter side, also measured in mm. The coefficient of friction, therefore, is dependent upon the relative roughness which is the ratio of absolute roughness and the internal pipe diameter k_s/d. Table 6.1 lists the surface roughness factors k_s for various materials.

Even with turbulent flow, the effect of fluid viscosity and friction at the boundary surface results in a film at the boundary wall which is known as

126 Laminar and turbulent flow

Table 6.1 Surface roughness factors for conduits

Material	k_s (mm)
Non-ferrous drawn tubing including plastics	0.0015
Black steel pipe	0.046
Aluminium ducting	0.05
Galvanised steel piping and ducting	0.15
Cast-iron pipe	0.20
Cement or plaster duct	0.25
Fair faced brick or concrete ducting	1.3
Rough brickwork ducting	5.0

Source: Reproduced from *CIBSE Guide* by permission of the Chartered Institution of Building Services Engineers.

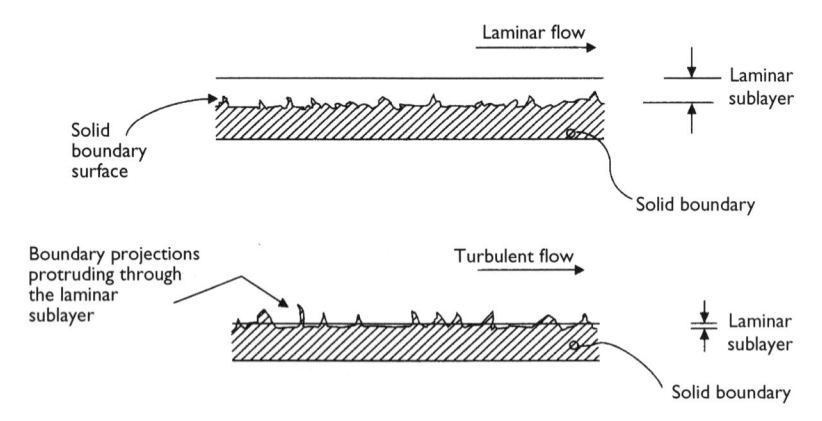

Figure 6.13 Surface roughness and the laminar sublayer (surface film).

the laminar sublayer. Under certain conditions, this may be sufficiently thick to obscure the high points on the boundary surface and flow will be as for a smooth pipe.

The film thickness reduces with increasing velocity and at some high value of Re rough projections protrude, increasing turbulence (refer to Figure 6.13) which in heat exchangers, for example, assists heat transfer.

6.6 Determination of the frictional coefficient in turbulent flow

A formula has been developed by Colebrook and White for the resolution of the frictional coefficient f in the D'Arcy equation 6.5

$$1/(f)^{0.5} = -4\log((k_s/3.7d) + 1.255/(Re)(f)^{0.5})$$

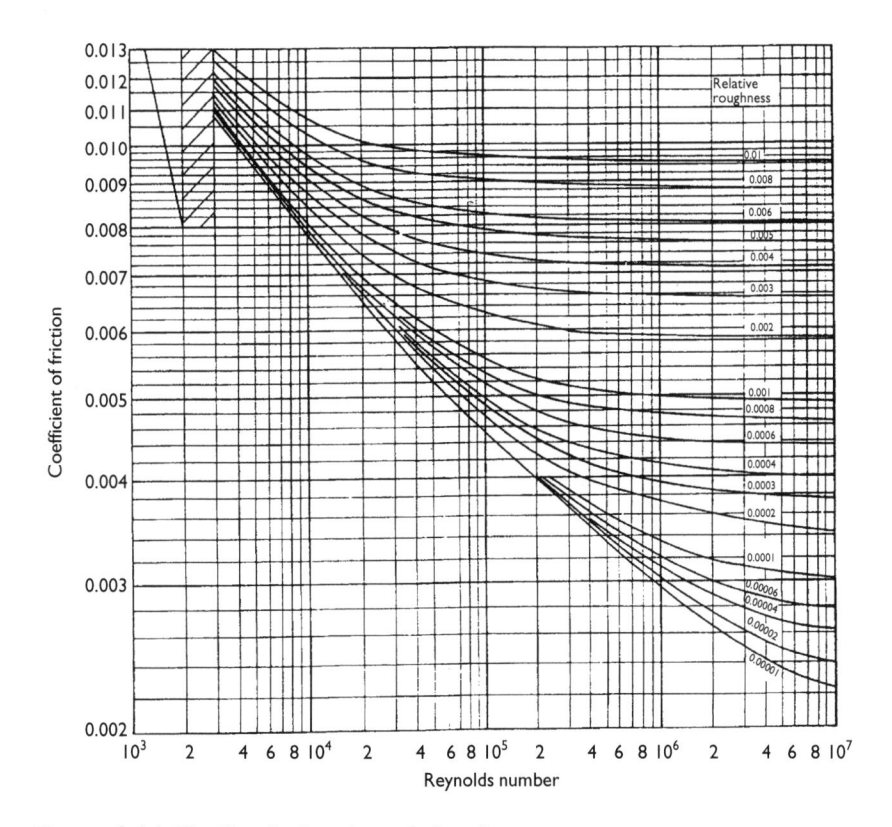

Figure 6.14 The Moody chart for turbulent flow.

Source: Reproduced from the *CIBSE Guide* (1986) by permission of the Chartered Institution of Building Services Engineers.

It can be seen that a simple solution to evaluate f using this formula requires a process of iteration. An alternative method of solution involves the use of the Moody chart of Poiseuille and Colebrook–White.

Figure 6.14 shows the Moody chart taken from section C of the *Guide* by the kind permission of *CIBSE*. You will see that after evaluating the Reynolds number and the relative roughness k_s/d for a particular application, the coefficient of friction f can be obtained by reading off the left-hand axis of the chart.

6.7 Solving problems

A number of problems and their solutions relating to laminar and turbulent flow are included in Chapter 7 that introduces and applies the theorem for the conservation of energy first proposed by Daniel Bernoulli in 1738. The problems considered here specifically relate to the two types of fluid flow.

128 Laminar and turbulent flow

You will see that fluids with low values of absolute viscosity have high values for the Reynolds number at relatively low mean velocities. It is important to remember that absolute viscosity varies with fluid temperature. For air it increases with increase in temperature but for water it decreases with temperature rise. Reference should be made to the tables of the *Thermodynamic and Transport Properties of Fluids* by Rogers and Mayhew.

Example 6.1 Calculate the pressure loss of water flowing in a pipe

A horizontal galvanised steel pipe is 80 mm nominal bore and 50 m in length. Determine the pressure loss sustained along the pipe if cold water flows at a mean velocity of 1.5 m/s. Determine also the specific pressure loss. Take water density as 1000 kg/m^3 and absolute viscosity as 0.001306 kg/ms.

Solution
From equation 6.6, Reynolds number $Re = \rho u d/\mu = 1000 \times 1.5 \times 0.08/0.001306 = 91\,884$. Flow is, therefore, turbulent. From Table 6.1 $k_s = 0.15$ and relative roughness $= k_s/d = 0.15/80 = 0.001875$. By locating the Reynolds number and the relative roughness on the Moody chart, Figure 6.14, the coefficient of friction $f = 0.0063$.

Adopting the D'Arcy equation for turbulent flow (equation 6.5)

$$dh = 4fLu^2/2gd = [4 \times 0.0063 \times 50 \times (1.5)^2]/[2 \times 9.81 \times 0.08]$$

$$= 1.806 \text{ m of water flowing}$$

$$dP = dh \cdot \rho \cdot g = 1.806 \times 1000 \times 9.81 = 17\,719 \text{ Pa}$$

SUMMARY FOR EXAMPLE 6.1

The specific pressure loss $dp = dP/L = 17\,719/50 = 354$ Pa/m. Specific pressure loss in pipe sizing is regulated by the maximum mean water velocity to avoid the generation of noise. For steel pipes above 50 mm nominal bore, this is 3 m/s or 4 m/s in long straight runs.

Referring to the pipe sizing tables in the *CIBSE Guide*, section C, and given a water velocity of 1.5 m/s and a calculated specific pressure drop of 354 Pa/m in 80 mm galvanised pipe at 10°C, the mass flow rate is interpolated as 7.22 kg/s.

Using the data in Example 6.1: the volume flow rate of water

$$Q = u \times [\pi \times d^2/4] = 1.5 \times [\pi \times (0.08)^2]/4 = 0.00754 \text{ m}^3/\text{s}$$

$$Q = 7.54 \text{ L/s}$$

Laminar and turbulent flow

and for cold water where

$$\rho = 1000 \text{ kg/m}^3$$

$$M = 7.54 \text{ kg/s}$$

This is close to the interpolated reading from the pipe sizing table of $M = 7.22$ kg/s.

It is helpful to find the maximum mean water velocity attainable in laminar flow here. This occurs when $Re < 2000$ and using equation 6.6

$$Re = \rho u d / \mu \quad \text{then} \quad u = Re \cdot \mu / \rho d = 2000 \times 0.001306 / 1000 \times 0.08$$
$$= 0.033 \text{ m/s}$$

This velocity is really too low for economic pipe sizing and hydraulic regulation. In fact, laminar flow rarely exists in systems of water distribution in building services.

Example 6.2 Determine the pumping power required and maximum laminar flow rate of oil in pipeline

Oil is pumped through a straight pipe 150 mm nominal bore and 80 m long. It discharges 10 m above the pump and neglecting all losses other than friction, determine:

(a) the power required to pump 16.67 kg/s of oil along the pipeline;
(b) the maximum mass transfer of oil that the pipe can transport in laminar flow and the pump power required.

Take oil density as 835 kg/m³ and viscosity as 0.12 kg/ms.

Solution (a)
From equation 6.6, $Re = \rho u d / \mu$, where mean velocity $u = M/\rho A = 4M/\rho \pi d^2$ m/s. Have a look at the units of the equation for u: (kg/s)/(kg/m³)(m²) = m/s. Thus,

$$u = 4 \times 16.67 / 835 \times \pi \times (0.15)^2 = 1.13 \text{ m/s}$$

Substituting into the Reynolds formula $Re = 835 \times 1.13 \times 0.15 / 0.12 = 1179$. Thus $Re < 2000$ and oil flow is laminar.

Adopting Poiseuille's equation 6.4, $dh = 32\mu u x / \rho g d^2$ m of oil flowing. Substituting

$$dh = 32 \times 0.12 \times 1.13 \times 80 / 835 \times 9.81 \times (0.15)^2 = 1.88 \text{ m}$$

130 Laminar and turbulent flow

Pump head required dh = viscous loss + elevation

$$dh = 1.88 + 10 = 11.88 \text{ m of oil flowing}$$

Now pump power,

$$P_w = M \cdot g \cdot dh \text{ Watts}$$

Having a look at the units of the terms on the right-hand side in the equation for P_w: (kg/s)(m/s^2)(m). The units can be rearranged thus: (kg m/s^2)(m/s). Now (kg m/s^2) are the basic SI units for force in Newtons N, where force = mass × acceleration. Thus the basic units for power

$$P_w = (\text{kg m/s}^2)(\text{m/s}) = \text{kg m}^2/\text{s}^3 \quad \text{or} \quad \text{N m/s} = \text{Watts}$$

Therefore substituting for pump power

$$P_w = 16.67 \times 9.81 \times 11.88 = 1943 \text{ W}$$

Note: The *dimensions* for power are ML^2T^{-3}.
Refer to Table A1.1 in Appendix 1.

Solution (b)
Laminar flow exists up to a maximum Reynolds number Re of 2000. From equation 6.6, $Re = \rho u d / \mu$ thus maximum mean velocity $u = \mu Re / \rho d$. Substituting:

$$u = 0.12 \times 2000 / 835 \times 0.15 = 1.916 \text{ m/s}$$

Since it was found in part (a) that $u = M/\rho A$, then $M = uA\rho$ kg/s. Then

$$M = 1.916 \times (\pi \times (0.15)^2/4) \times 835 = 28.274 \text{ kg/s}$$

From equation 6.4,

$$dh = 32\mu u x / \rho g d^2 = 32 \times 0.12 \times 1.916 \times 80 / 835 \times 9.81 \times (0.15)^2$$
$$dh = 3.19 \text{ m of oil flowing}$$

Pump head required dh = viscous loss + elevation (static lift)

$$dh = 3.19 + 10 = 13.19 \text{ m of oil flowing}$$

Pump power

$$P_w = M \cdot g \cdot dh = 28.274 \times 9.81 \times 13.19 = 3658 \text{ W}$$

SUMMARY AND CONCLUSIONS FOR EXAMPLE 6.2

Mass transfer M (kg/s)	Pipe diameter d (mm)	Reynolds no. Re	Mean velocity u (m/s)	Pump power P_W (W)
16.67	150	1179	1.13	1943
28.274	150	2000	1.916	3658

- The pump power is the output power and does not account for pump efficiency.
- An increase in mass transfer of $(28.274 - 16.67)/16.67 = 70\%$, results in an increase in pump power required of $(3658 - 1943)/1943 = 88\%$.
- Fluid viscosity is temperature dependent. The viscosity of fuel oils is particularly sensitive to temperature and medium and heavy fuel oils require heating before pumping can begin. The pipeline will also need to be well insulated and, depending upon its viscosity, may require tracing to maintain the temperature and hence satisfactory oil flow.
- Laminar flow as well as turbulent flow can occur in systems of oil distribution.

Example 6.3 Show that maximum velocity is twice the mean in laminar flow through a pipe

Given that the velocity at radius r for laminar flow is expressed as:

$$u = dP(R^2 - r^2)/4\mu L$$

where dP is the pressure drop over length L and R is the pipe inside radius. Show that the maximum velocity is twice the mean velocity.

Solution
From Poiseuille's equation 6.4, $dh = 32\mu u L/\rho g d^2$ m of oil flowing. Rearranging in terms of mean velocity u:

$$u = dh \cdot \rho g d^2/32\mu L \text{ m/s}$$

Since $dP = dh \cdot \rho g$, mean velocity

$$u = dP \cdot d^2/32\mu L \text{ m/s}$$

Actual velocity at radius r,

$$u = dP(R^2 - r^2)/4\mu L \text{ m/s}$$

132 Laminar and turbulent flow

Maximum velocity will occur at the pipe centreline when $r=0$. Thus maximum velocity

$$u = dP \cdot R^2/4\mu L$$

Since maximum velocity = twice the mean velocity,

$$dP \cdot R^2/4\mu L = 2(dP \cdot d^2/32\mu L)$$

Then since $R^2 = (d/2)^2$, $dP(d/2)^2/4\mu L = dP \cdot d^2/16\mu L$. Therefore

$$dP \cdot d^2/16\mu L = dP \cdot d^2/16\mu L$$

Thus maximum velocity = twice the mean velocity in laminar flow.

Example 6.4 Calculate the diameter of the rising main that serves two high level tanks

Two cold water tanks, each with 4500 L capacity, are refilled every 2 h. The vertical height of the water main is 26 m and its horizontal distance from the water utility's main is 9 m. If the available pressure is 300 kPa during peak demand, calculate the diameter of the rising main.

Data: Pressure required at the ball valve is 30 kPa, make an allowance for pipe fittings of 10% on straight pipe, assume initially that the coefficient of friction f is 0.007, the viscosity of cold water is 0.001306 kg/ms and the water density is 1000 kg/m³.

Solution

Figure 6.15 shows the arrangement in elevation.

Flow rate required in the rising main = $4500 \times 2/2 \times 3600 = 1.25$ L/s

$= 0.00125$ m³/s

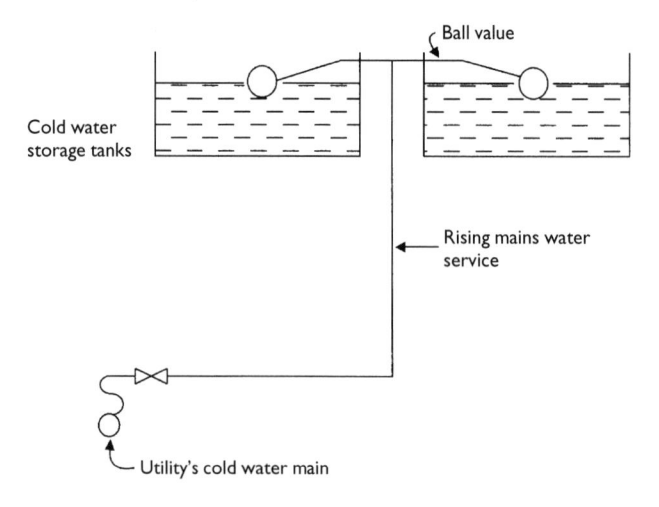

Figure 6.15 Example 6.4.

Mains pressure available for pipe sizing

$$= 300 - \text{static lift pressure} - \text{pressure at ball valve}$$
$$= 300 - h\rho g - 30$$
$$= 300 - (26 \times 1000 \times 9.81/1000) - 30$$
$$dP = 300 - 255 - 30 = 15 \text{ kPa}$$

Since $dP = dh \cdot \rho g$

$$dh = dP/\rho g = 15\,000/1000 \times 9.81 = 1.53 \text{ m of water}$$

Total equivalent length of pipe and fittings $= (26 + 9)1.1 = 38.5$ m.

Initially assuming turbulent flow and adopting Box's formula, equation 6.8 becomes $Q = (3dh \cdot d^5/fL)^{0.5}$; and rearranging in terms of pipe diameter d:

$$d = (fLQ^2/3dh)^{1/5}$$

Substituting:

$$d = \{[0.007 \times 38.5 \times (0.00125)^2]/(3 \times 1.53)\}^{1/5} = (9.1742 \times 10^{-8})^{1/5}$$
$$d = 0.039 \text{ m}$$

Thus standard pipe diameter $d = 40$ mm.

It is now necessary to check that water flow is turbulent and to verify the value of the coefficient of friction f. Since $Q = uA = u \times [\pi d^2/4]$, mean velocity

$$u = 4Q/\pi d^2 = [4 \times 0.00125]/[\pi(0.04)^2] = 1 \text{ m/s}$$

From equation 6.6

$$Re = \rho u d/\mu = [1000 \times 1 \times 0.04]/[0.001306] = 30\,628$$

Thus flow is turbulent and adopting D'Arcy's equation 6.5, $dh = 4fLu^2/2gd$. Rearranging in terms of frictional coefficient

$$f = 2dh \cdot gd/4Lu^2$$

Substituting:

$$f = [2 \times 1.53 \times 9.81 \times 0.04]/[4 \times 38.5 \times 1^2]$$
$$f = 0.0078$$

134 Laminar and turbulent flow

The frictional coefficient used in the solution was $f = 0.007$. The effect on the pipe diameter can be shown by recalculation where $d = (fLQ^2/3dh)^{1/5}$; and substituting using $f = 0.0078$ this time,

$$d = [0.0078 \times 38.5 \times (0.00125)^2/(3 \times 1.53)]^{1/5}$$

pipe diameter, $d = 0.04$ m $= 40$ mm

SUMMARY FOR EXAMPLE 6.4

Clearly the small error in the initial value for the coefficient of friction is insignificant here. The solution to this problem can be achieved by applying Bernoull's theorem for the conservation of energy. This is discussed in Chapter 7. If the theorem is applied here, taking section 1 to be at incoming mains level and section 2 to be at tank level:

$$Z_1 + (P_1/\rho g) + (u_1^2/2g) = Z_2 + (P_2/\rho g) + (u_2^2/2g) + \text{losses}$$

Since $u_1 = u_2$,

$$((P_1 - P_2)/\rho g) + Z_1 - Z_2 = \text{losses}$$

Substituting:

$$((300\,000 - 30\,000)/\rho g) + 0 - 26 = fLQ^2/3d^5$$

Thus

$$1.53 = 0.007 \times 38.5 \times (0.00125)^2/3d^5$$

from which pipe diameter d can be evaluated and $d = 40$ mm. You should now confirm this solution.

Example 6.5 Determine the static pressure loss of air flowing in a duct

Air at 27°C flows at a mean velocity of 5 m/s in a 30 m straight length of galvanised sheet steel duct 400 mm diameter. Determine the static pressure loss along the duct due to air flow. Data taken from the tables of *Thermodynamic and Transport of Fluids* for dry air at 300 K: $\mu = 0.00001846$ kg/m·s, $\rho = 1.177$ kg/m^3.

Solution

The type of flow can be identified from the Reynolds number, equation 6.6

$$Re = \rho u d/\mu = [1.177 \times 5 \times 0.4]/[0.00001846] = 127\,935$$

Since $Re > 2000$ flow is turbulent and D'Arcy's equation 6.5 can be used to find the pressure loss in the duct. However, it is first of all necessary to find the coefficient of friction f in the D'Arcy equation and this can be done by using the Moody chart, Figure 6.14.

From Table 6.1, the absolute roughness of the duct wall is 0.15 mm and relative roughness = $k_s/d = 0.15/400 = 0.000375$. Using the calculated value of Re and relative roughness, the coefficient of friction from the Moody chart is $f = 0.0047$. Substituting into the D'Arcy equation

$$dh = [4 \times 0.0047 \times 30 \times 5^2]/[2 \times 9.81 \times 0.4]$$

$$dh = 1.8 \text{ m of air flowing}$$

Since $dP = dh \cdot \rho g$, pressure loss

$$dP = 1.8 \times 1.177 \times 9.81 = 20.8 \text{ Pa}$$

SUMMARY FOR EXAMPLE 6.5

- The specific pressure drop $dp = dP/L = 20.8/30 = 0.7$ Pa/m. A typical rate of pressure drop in straight ducts for low pressure ventilation and air conditioning systems in which the maximum mean air velocity is around 5 m/s is around 1.0 Pa/m.
- Two pressures are present in a system of air flow namely static pressure and velocity pressure. This is discussed more fully in Chapters 5 and 7.
- It is assumed for the purposes of duct sizing that air behaves as an incompressible fluid. This is not necessarily the case at the prime mover or fan where its operating characteristics can show the effects of compression. You should refer to fan manufacturers' literature here.
- It is helpful to consider the maximum mean air velocity attainable in laminar flow in the duct. For laminar flow $Re < 2000$, thus using equation 6.6,

    $$2000 = 1.177 \times u \times 0.4/0.00001846$$

    from which $u = 0.08$ m/s. This air velocity is too low for ductwork design; it is therefore invariably in the turbulent region.

6.8 Chapter closure

This chapter has provided you with the underpinning knowledge of the two models of fluid flow, namely laminar and turbulent, relating to the flow of water, oil and air in pipes and ducts. It provides a methodology for identifying the type of flow in a system and procedures for solving some problems. The text also defines the pressure losses and mean fluid velocities which one may find in systems conveying oil, air and water. The characteristics of laminar and turbulent flow are discussed in the context of containment in the solid boundaries of the pipe or duct.

Further work is investigated in Chapters 5 and 7. Chapter 7 includes partial flow in pipes and flow in open channels. Part of Appendix 2 focuses on the dimensionless numbers used here and elsewhere in the book.

Chapter 7

Mass transfer of fluids in pipes, ducts and channels

Nomenclature

a	area of cross section (m²)
A	surface area (m²)
b	breadth (mm, m)
C	Bernoulli constant, Chezy constant
C_c	coefficient of contraction
d	diameter, depth (mm, m)
dh	head loss in metres of fluid flowing
dP	pressure loss (Pa, kPa)
dp	specific pressure loss (Pa/m)
f	coefficient of friction
g	gravitational acceleration at sea level 9.81 m/s²
H	energy (Nm, J)
i	hydraulic gradient (m/m)
K	constant
k	velocity head/pressure loss factor
k_s	surface roughness (mm)
L	length (m)
m	mass (kg), hydraulic mean diameter/depth (m)
M	mass transfer (kg/s)
n	roughness coefficient
P	impermeability
P	pressure (Pa, kPa)
Q	volume flow rate (m³/s, L/s)
R	rainfall intensity (mm/h)
Re	Reynold number
S	ratio of densities
u	mean velocity of flow (m/s)
x	displacement of measuring fluid (mm, m)
Z	vertical height in relation to a datum (m)
μ	viscosity (kg/ms)
ρ	density (kg/m³)

7.1 Introduction

This chapter focuses upon the determination of mass transfer of fluids subject to a prime mover and to gravity. This forms a significant part of the design of heating, ventilating and air conditioning systems and hot and cold water supply. It also focuses on pressure loss resulting from frictional flow and on the hydraulic gradient.

7.2 Solutions to problems in frictionless flow

In Chapter 5, Bernoulli's conservation of energy at a point for a moving fluid or a stationary fluid having potential energy was introduced and stated that:

potential energy + pressure energy + kinetic energy = total energy
$$= \text{a constant}$$

Thus $Z + (P/\rho g) + (u^2/2g) = C$ in metres of fluid flowing.

In pressure units, each of the terms in the Bernoulli theorem must be multiplied by ρ and g; thus

$$(Z \cdot \rho g) + P + (\rho \cdot u^2/2) = C \, \text{Pa}$$

In energy units of Joules or Nm, the Bernoulli theorem in metres of fluid flowing must be multiplied by m and g; thus

$$(Z \cdot mg) + (P \cdot m/\rho) + (m \cdot u^2/2) = C \, \text{J or Nm}$$

The dimensions of these terms can be checked to ensure integrity. This process is considered in detail in Appendix 1.

Chapter 5 also introduced the Bernoulli statement that the total energy of a moving fluid at one point in a system is equal to the total energy of that fluid at some point downstream. The following example illustrates this statement.

Example 7.1 Find the maximum length of a pump suction pipe set to a gradient

The suction pipe of a pump rises from a ground storage tank at a slope of 1 in 7 and cold water is conveyed at 1.8 m/s. If dissolved air is released when the pressure in the pipe falls to more than 50 kPa below atmospheric pressure, find the maximum practicable length of pipe ignoring the effects of friction. Assume that the water in the tank is at rest.

Solution
Figure 7.1 shows a diagram of the system in elevation. Applying the Bernoulli equation for the conservation of energy at points 1 and 2 in the system

138 Mass transfer of fluids

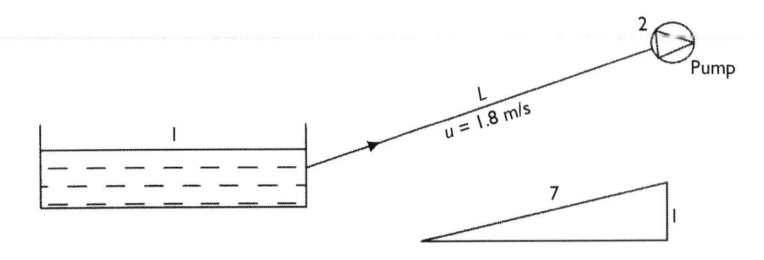

Figure 7.1 Example 7.1.

(see Section 5.3) and taking atmospheric pressure as 101 325 Pa:

$$Z_1 + (P_1/\rho g) + (u_1^2/2g) = Z_2 + (P_2/\rho g) + (u_2^2/2g)$$

Thus by substitution

$$Z_1 + (101\,325/\rho g) + 0 = Z_2 + [(101\,325 - 50\,000)/\rho g] + (1.8)^2/2g$$

Rearranging

$$Z_2 - Z_1 = [(101\,325 - 51\,325)/\rho g] - (1.8)^2/2g$$

$$dZ = 5.097 - 0.165 = 4.932\,\text{m}$$

This is the vertical length of the allowable rise from the ground storage tank. Since the pump is located at the upper point it is called suction lift.

The maximum length of a suction pipe for a gradient of 1 in 7, by Pythagoras is

$$L = [(4.932 \times 7)^2 + (4.932)^2]^{0.5}$$

$$L = 34.9\,\text{m}$$

SUMMARY TO EXAMPLE 7.1

- As the gradient increases, the maximum practicable length decreases; and for a gradient of 1 in 3, practical length $L = 15.6$ m. You should confirm that this is so. For a vertical pipe, practical length $L = 4.93$ m. These practical lengths relate to a subatmospheric pressure in the suction pipe at section 2 of 51 kPa, hence the term suction lift.
- The theoretical maximum vertical length (suction lift) for cold water in this pipe will occur when atmospheric pressure in the suction pipe at section 2 is zero. It can be obtained by applying the Bernoulli equation, thus:

$$Z_1 + (P_1/\rho g) + (u_1^2/2g) = Z_2 + (P_2/\rho g) + (u_2^2/2g)$$

substituting

$$Z_1 + (101\,325/1000 \cdot g) + 0 = Z_2 + 0 + (1.8)^2/2g$$

re-arranging

$$Z_2 - Z_1 = 10.33 - 0.165 = 10.164\,m$$

and maximum theoretical vertical lift $L = 10.16$ m.
This assumes a mean water velocity in the suction pipe of 1.8 m/s.

- This amount of suction lift is impossible to achieve. At zero atmospheric pressure within the suction pipe, water vaporises at 0°C and it will therefore evaporate before reaching the impeller of the pump which is not designed anyway to handle vapour.

As the pump generates negative pressure in the suction pipe, water will be drawn up to a point where its absolute pressure corresponds to its saturation temperature and partial evaporation occurs. Priming the pump will not assist it to achieve a suction lift of this magnitude.

The maximum practical vertical suction lift for cold water is about 5 m. Since the water pressure in the pipe is subatmospheric the pipe must not be made from collapsible materials, such as canvas. If water must be pumped from a point lower than 5 m, a submersible pump is employed and located in the water contained in the tank or well.

Example 7.2 Determine the mass transfer of water passing through a syphon pipe

A 75 mm bore syphon pipe rises 1.8 m from the surface of water in a tank and drops to a point 3.6 m below the water level where it discharges water to atmosphere. Ignoring the effects of friction, determine the discharge rate in L/s and the absolute pressure of the water at the crest of the syphon. Take atmospheric pressure as equivalent to 10 m of water and water density as 1000 kg/m³.

Solution
Figure 7.2 shows the arrangement in elevation. Adopting the Bernoulli equation for frictionless flow at points 1 and 3, and taking point 3 as datum:

$$Z_1 + P_1/\rho g + (u_1^2/2g) = Z_3 + (P_3/\rho g) + (u_3^2/2g)$$

140 Mass transfer of fluids

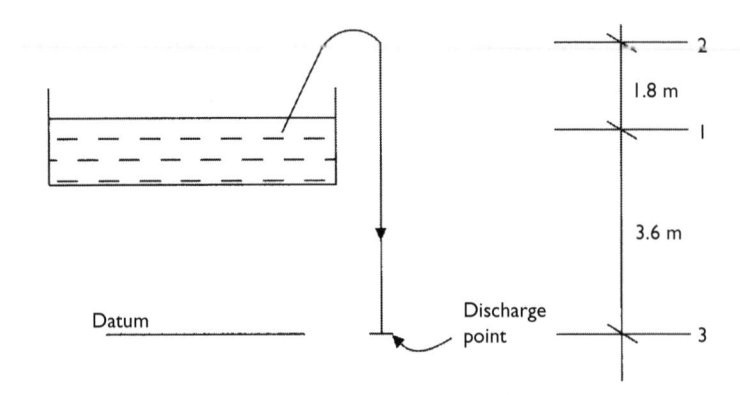

Figure 7.2 Example 7.2.

where $Z_1 = 3.6$ m, $Z_3 = 0$, $u_1 = 0$, $P_1 = P_3 = 10$ m. Substituting

$$3.6 + 10 + 0 = 0 + 10 + (u_3^2/2g)$$

from which

$$u_3^2 = 3.6 \times 2g = 70.632$$

and

$$u_3 = 8.4 \text{ m/s}$$

Rate of discharge

$$Q = u \times (\pi d^2/4) = 8.4 \times \pi \times 0.075^2/4 = 0.037 \text{ m}^3/\text{s}$$
$$Q = 37 \text{ L/s}$$

Equating points 1 and 2 and keeping point 3 as datum:

$$Z_1 + (P_1/\rho g) + (u_1^2/2g) = Z_2 + (P_2/\rho g) + (u_2^2/2g)$$

For uniform flow $u_1 = u_3 = 8.4$ m/s. Substituting

$$3.6 + 10 + 0 = (3.6 + 1.8) + (P_2/\rho g) + (8.4)^2/2g$$

Rearranging

$$P_2 = [(13.6 - 5.4) - (8.4)^2/2g]\rho g$$

From which

$$P_2 = (13.6 - 5.4 - 3.6)1000g = 45\,126 \text{ Pa}$$

Thus the absolute water pressure at the crest of the siphon = 45 kPa. Alternatively, subatmospheric pressure at this point is 45 kPa. From the summary to Example 7.1 you can see that the effect of the low water pressure at the crest of the siphon may cause the water to separate. It is likely therefore that the discharge will be erratic.

Example 7.3 Calculate the gauge pressure at a hydrant valve when it is shut and when it is open

A 65 mm bore fire hydrant is fed from a water tank located 37 m vertically above. A pressure gauge and stop valve are fitted at the hydrant and with the valve fully open water flows at 26 L/s. Determine the gauge pressure reading

(a) with the hydrant valve fully open;
(b) with the valve shut.

Solution (a)
Figure 7.3 shows the system in elevation. Adopting the Bernoulli equation for the conservation of energy at points 1 and 2

$$Z_1 + (P_1/\rho g) + (u_1^2/2g) = Z_2 + (P_2/\rho g) + (u_2^2/2g)$$

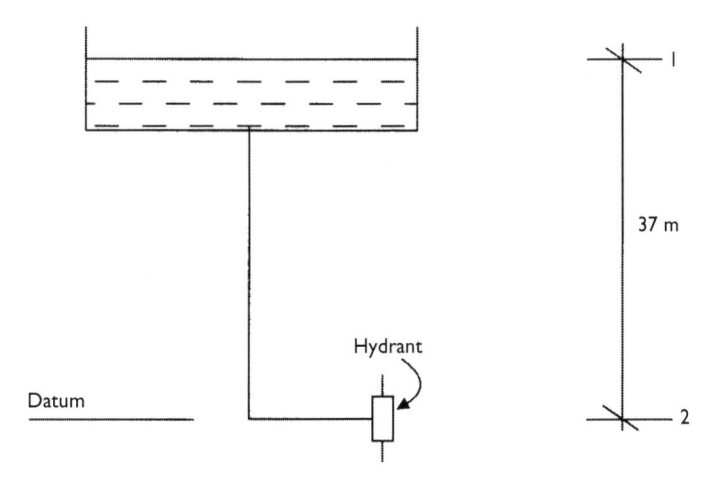

Figure 7.3 Example 7.3.

where water velocity in the tank $u_1 = 0$. Water velocity in the pipe

$$u_2 = Q/a = 0.026 \times (4/\pi d^2) = 0.026 \times 4/\pi \times (0.065)^2 = 7.835 \text{ m/s}.$$

Pressure P_1 at the tank is atmospheric and therefore the gauge pressure is zero. Substituting

$$37 + 0 + 0 = 0 + P_2/\rho g + [(7.835)^2/2g]$$

from which

$$P_2 = (37 - 3.13) \times 1000 \times 9.81 = 332\,226 \text{ Pa}$$

and

$$P_2 = 332 \text{ kPa}$$

Solution (b)
With no flow $u_1 = u_2 = 0$. Substituting in the Bernoulli equation:

$$37 + 0 + 0 = 0 + (P_2/\rho g) + 0$$

from which

$$P_2 = 37 \times 1000 \times 9.81 = 362\,970 \text{ Pa}$$

and

$$P_2 = 363 \text{ kPa}$$

Example 7.4 Find the diameter at the larger end of a transformation piece through which air is flowing

Air at 30°C flows in a horizontal circular duct and at a section A mean velocity is 7 m/s and the static pressure is 300 Pa. If the duct expands gradually from 380 mm at point A find the duct diameter downstream at point B where the static pressure is registered as 320 Pa. Take standard air density at 20°C as 1.2 kg/m³.

Solution
Figure 7.4 shows the system in elevation. Adopting the Bernoulli equation for frictionless flow:

$$Z_a + (P_a/\rho g) + (u_a^2/2g) = Z_b + (P_b/\rho g) + (u_b^2/2g)$$

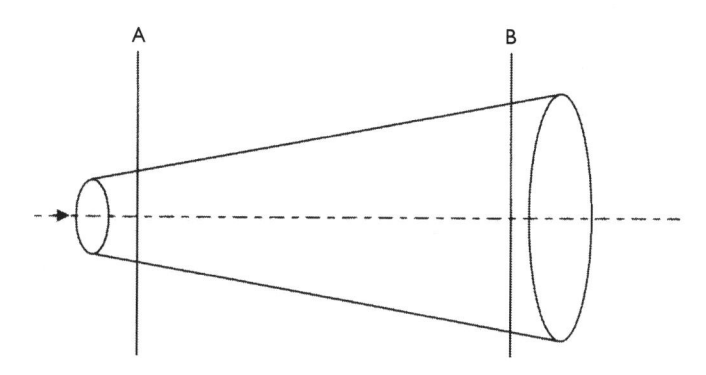

Figure 7.4 Example 7.4.

As the duct is horizontal $Z_a = Z_b$. Ignoring the effect of pressure variations on air density, for air at 30°C

$$\rho_2 = \rho_1(T_1/T_2) = 1.2(273 + 20)/(273 + 30) = 1.16 \text{ kg/m}^3$$

Thus

$$0 + (300/1.16 \times 9.81) + (7^2/2g) = 0 + (320/1.16 \times 9.81) + (u_b^2/2g)$$

and

$$26.36 + 2.5 = 28.12 + (u_b^2/2g)$$

from which

$$u_b^2 = 0.74 \times 2 \times 9.81 = 14.5$$

so

$$u_b = 3.81 \text{ m/s}$$

Volume flow rate

$$Q_a = u_a \times (\pi d a^2/4) = 7 \times \pi \times 0.38^2/4 = 0.794 \text{ m}^3/\text{s}$$

For steady flow $Q_a = Q_b$. Thus

$$0.794 = 3.81 \times \pi \times d_b^2/4$$

from which

$$d_b^2 = 4 \times 0.794/3.81 \times \pi = 0.265$$

144 Mass transfer of fluids

and

$$d_b = 0.515 \text{ m} = 515 \text{ mm diameter}$$

SUMMARY TO EXAMPLE 7.4
You would have seen that as the duct transformation piece is an expansion, mean air velocity decreases and static pressure increases. This is known as static regain. Refer to Example 7.13.

7.3 Frictional flow in flooded pipes and ducts

Consider the following example as an introduction to frictional flow.

Example 7.5 Determine the power available at the nozzle of an open fire hydrant valve

A jet of water issuing at a velocity of 22.5 m/s is discharged through a fire hydrant nozzle having a diameter of 75 mm.

(a) Determine the power of the issuing jet if the nozzle is supplied from a reservoir 30 m vertically above.
(b) What is the loss of head in the pipeline and nozzle?
(c) What is the efficiency of power transmission?

Take the density of water as 1000 kg/m^3.

Solution(a)
The energy at the nozzle $= Z + (P/\rho g) + (u^2/2g)$ metres of water flowing. At the nozzle the potential energy is zero and the pressure energy is converted to kinetic energy, thus:

$$\text{Energy at the nozzle} = 0 + 0 + u^2/2g = 22.5^2/2g = 25.8 \text{ m of water}$$
$$\text{Power} = Q \cdot dP$$

where

$$Q = u \cdot a = 22.5 \times [\pi(0.075^2)/4] = 0.0994 \text{ m}^3/\text{s}$$

and

$$dP = dh \cdot \rho \cdot g = 25.8 \times 1000 \times 9.81 = 253\,098 \text{ Pa}$$

thus

$$\text{power} = 0.0994 \times 253\,098 = 25\,158 \text{ W}$$

and

power = 25.2 kW.

Solution (b)
The Bernoulli equation is easily adapted to frictional flow and applying it here to sections A and B, in Figure 7.5,

$$Z_a + (P_a/\rho g) + (u_a^2/2g) = Z_b + (P_b/\rho g) + (u_b^2/2g) + \text{loss}$$

Substituting

$$30 + 0 + 0 = 0 + 0 + (u_b^2/2g) + \text{loss}$$

from which frictional loss = 30 − 25.8 = 4.2 m of water.

Solution (c)
The efficiency of power transmission = energy at the nozzle/energy at the reservoir. The energy at the nozzle is in the form of kinetic energy and the energy in the water stored in the high level reservoir is in the form of potential energy. Thus efficiency = 25.8/30 = 0.86 = 86%.

SUMMARY TO EXAMPLE 7.5
You will notice the significant value of the power transmission. It confirms the reason why more than one fireman may be required to hold the nozzle steady when it is connected by canvas hose to the hydrant. The losses due to friction will account for the loss at the exit from the reservoir, the loss in

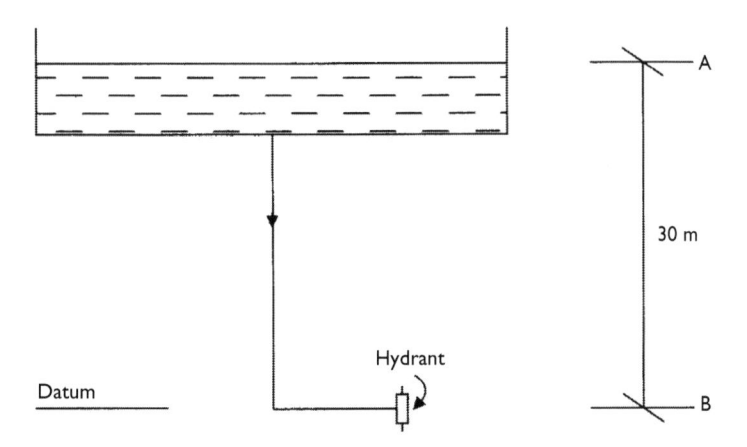

Figure 7.5 Example 7.5.

straight pipe and the loss in pipe fittings. It will be shown below that there is no shock loss at the nozzle.

Frictional losses in pipes, ducts and fittings

Frictional losses in pipe and duct systems may therefore include the following:

- Shock losses.
- Losses in the straight pipe or duct.
- Losses in fittings including bends, tees, valves, volume control dampers, etc.
- Manufacturers of items of appropriate plant will provide the loss due to friction at given flow rates.

Shock loss usually occurs at sudden enlargements and sudden contractions. The entry to and exit from a large vessel such as the flow and return connections on the secondary side of a hot water service calorifier provides one example. In the case of air flow, shock loss occurs across a supply air diffuser and a return air grille.

For a sudden enlargement shock loss, $dh = (u_1 - u_2)^2/2g$ metres of fluid flowing. A special case occurs when water discharges into a large tank of water or air is discharged into a room. In these cases, u_2 approaches zero and therefore shock loss

$$dh = u_1^2/2g$$

A further special case occurs when water discharges into air. In this example, $u_1 = u_2$ and therefore shock loss $dh = $ zero.

- For a sudden contraction, shock loss $dh = 0.5(u_2^2/2g)$ metres of fluid flowing.
- For frictional losses in straight pipes and ducts in which the fluid flow is turbulent the D'Arcy equation applies and $dh = 4fLu^2/2gd$ metres of fluid flowing.
- Frictional losses in fittings are based upon the velocity head loss and $dh = k \cdot (u^2/2g)$ metres of fluid flowing; $k = $ the velocity head loss factor for the fitting. Since it is dimensionless it has the same value as the velocity pressure loss factor k. Typical values of k are given for a variety of pipe and duct fittings in section C4 of the *CIBSE Guide* where it is identified as Greek letter zeta [ζ].

The shock loss for a sudden contraction given above is dependent upon the coefficient of contraction C_c at the fluid's vena contracta downstream of

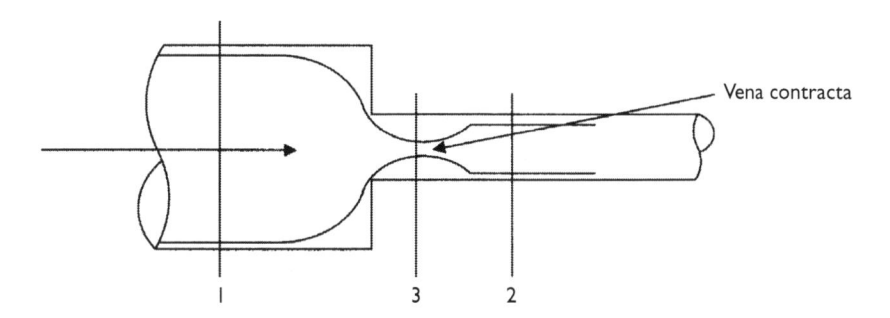

Figure 7.6 The sudden contraction.

the fitting. See Figure 7.6.

$$C_c = a_3/a_2 \quad \text{and} \quad dh = u_2^2/2g[(1/C_c) - 1]^2$$

The term $[(1/C_c) - 1]^2$ reduces to 0.5 when $C_c = 0.585$.

It is usually assumed that $C_c = 0.585$ and hence shock loss $dh = 0.5(u_2^2/2g)$ is taken for most sudden contractions due to the difficulty in determining the coefficient of contraction which has to be done by experiment for each fitting. u_1 refers to the mean velocity of the fluid upstream of the fitting and u_2 refers to the mean velocity of the fluid downstream of the fitting.

- A final special case relates to the use of the bellmouth at the entry into or exit from a large vessel in the case of water or exit from/entry into a room in the case of air flow.

The bellmouth replaces the sharp edge of the entry or exit point with a radiused 'edge'.

This has the effect of reducing the shock loss to zero. There now follows some examples which apply the rational formulae introduced in this section.

Example 7.6 Calculate the shock loss as water flows through a sudden enlargement

A horizontal pipe transporting water at 12 L/s suddenly increases from 100 mm to 200 mm in bore. Determine the shock loss in metres of water flowing, in units of pressure and units of energy. Take the density of water as 1000 kg/m^3.

Solution
For a sudden enlargement $dh = (u_1 - u_2)^2/2g$ metres of water flowing:

$$u_1 = Q/a = 0.012 \times 4/\pi \times (0.1)^2 = 1.528 \text{ m/s}$$
$$u_2 = 0.012 \times 4/\pi \times (0.2)^2 = 0.382 \text{ m/s}$$

Substituting

$$dh = (1.528 - 0.382)^2/2g = 0.067 \text{ m of water flowing}$$

In units of pressure

$$dP = dh \cdot \rho \cdot g = 0.067 \times 1000 \times 9.81 = 657 \text{ Pa}$$

In units of energy

$$dH = dh \cdot m \cdot g = 0.067 \times 12 \times 9.81 = 7.89 \text{ Nm (J)}$$

Example 7.7 Determine the shock loss and mass transfer as oil passes through a sudden enlargement

A horizontal pipe carrying oil suddenly increases from 80 to 150 mm in bore. The fluid displacement in a differential manometer connected on either side of the enlargement is 18 mm. Determine

(a) the shock loss
(b) the flow rate of oil and
(c) analyse the relationship between the manometer reading, static regain and shock loss.

Take the density of oil as 935 kg/m^3 and that for the measuring fluid as 13 600 kg/m^3.

Note: For a sudden enlargement in a pipe or duct the fluid being transported suffers a drop in velocity where $u_1 > u_2$ hence there is a loss of velocity pressure. This produces a corresponding rise in static pressure. So $P_2 > P_1$. The manometer reading will show the algebraic sum of the shock loss and static gain. This registers as a negative displacement of the measuring fluid. That is to say, the liquid level in the manometer limb connection downstream is lower than that in the upstream limb. See Figure 5.4.

Solution (a)
Section 1 is upstream and section 2 is downstream of the sudden enlargement. For a sudden enlargement $u_1 > u_2$ therefore $P_2 > P_1$ and the loss in velocity pressure will be equal to the gain in static pressure. This gain is referred to as static regain.

From Chapter 5, the equivalent displacement of oil

$$dh = x(S-1) = 0.018[(13\,600/935) - 1]$$

from which $dh = 0.244$ m of oil flowing. For steady flow, $Q_1 = Q_2$ and therefore $u_1 \cdot a_1 = u_2 \cdot a_2$ from which $u_2 = u_1 \times a_1/a_2$ thus $u_2 = u_1 \times (d_1/d_2)^2$. Substituting, $u_2 = u_1 \times (80/150)^2 = 0.285\, u_1$.

Adopting the Bernoulli equation for frictional flow in which $Z_1 = Z_2$ and the frictional loss for a sudden enlargement

$$dh = (u_1 - u_2)^2/2g$$

thus

$$(P_1/\rho g) + (u_1^2/2g) = (P_2/\rho g) + (u_2^2/2g) + (u_1 - u_2)^2/2g$$

rearranging

$$(P_2 - P_1)/\rho g = (u_1^2 - u_2^2)2g - (u_1 - u_2)^2/2g$$

Remember that for an enlargement there is a gain in static head, known as static regain and a loss in velocity head, thus $P_2 > P_1$ and $u_1 > u_2$. Expanding the right-hand side

$$(P_2 - P_1)/\rho g = [(u_1^2 - u_2^2) - (u_1^2 - 2u_1 \cdot u_2 + u_2^2)]/2g$$

thus

$$(P_2 - P_1)/\rho g = 2u_1 \cdot u_2 - 2u_2^2/2g$$
$$= 2u_2(u_1 - u_2)/2g$$

Substituting:

$$0.244 = [2 \times 0.285u_1(u_1 - 0.285u_1)]/2g$$
$$0.244 \times 2g = (0.40755u_1^2)$$
$$u_1^2 = (0.244 \times 2g)/(0.40755) = 11.7465$$

and

$$u_1 = 3.427 \text{ m/s}$$

Shock loss

$$dh = [3.427 - (0.285 \times 3.427)]^2/2g = 0.306 \text{ m of oil flowing}$$

Solution (b)
Flow rate

$$Q = u \times a = 3.427 \times \pi \times (0.08)^2/4 = 0.0172 \text{ m}^3/\text{s} = 17.2 \text{ L/s}$$

Solution (c)
From Bernoulli's equation, velocity head $= u^2/2g$ metres of fluid flowing.

$$\text{The loss in velocity head} = (u_1^2 - u_2^2)/2g$$

where $u_2 = 0.285 u_1 = 0.285 \times 3.427 = 0.977$ m/s.

$$\text{So loss in velocity head} = [(3.427)^2 - (0.977)^2]/2g$$
$$= 0.55 \text{ m of oil flowing}$$

Since loss in velocity head = gain in static head, the increase in static head = 0.55 m of oil flowing. This is the gross static regain.

The manometer will register the net increase in pressure across the sudden enlargement; so it will read the algebraic sum of the regain in static pressure and the shock loss. This will be $[+0.55 - 0.306] = 0.244$ m which is the equivalent manometer reading above in metres of oil flowing as calculated in part (a). Note that the manometer therefore is reading an *increase* pressure (negative displacement) equivalent to 18 mm of measuring fluid in the instrument.

Example 7.8 Calculate the shock loss and mass transfer as water passes through a sudden contraction

A horizontal pipe carrying water suddenly contracts from 300 mm to 100 mm bore. A differential manometer is connected on either side of the sudden contraction and the pressure drop is 8.34 kPa. Determine

(a) the shock loss;
(b) the mass transfer of water and
(c) analyse the relationship between the manometer reading, shock loss and loss in static pressure.

Take the density of water as 1000 kg/m^3.

Solution (a)
Adopting the Bernoulli equation for frictional flow, and given $Z_1 = Z_2$

$$(P_1/\rho g) + (u_1^2/2g) = (P_2/\rho g) + (u_2^2/2g) + 0.5 u_2^2/2g$$

Refer to Figure 7.6 and rearrange, having in mind that $P_1 > P_2$ and therefore $u_2 > u_1$

$$(P_1 - P_2)/\rho g = [(u_2^2 - u_1^2)/2g] + 0.5 u_2^2/2g$$

Now, $u_2 = u_1(d_1/d_2)^2 = u_1(300/100)^2 = 9u_1$. Substituting

$$8340/1000g = [(9u_1)^2 - u_1^2]/2g + 0.5 \times (9u_1)^2/2g$$
$$0.85 \times 2g = 80u_1^2 + 40.5u_1^2$$
$$16.68 = 120.5u_1^2$$

from which $u_1 = 0.372$ m/s.
 Shock loss

$$dh = 0.5(9 \times 0.372)^2/2g = 0.286 \text{ m of water flowing}$$
$$dP = dh \cdot \rho \cdot g = 0.286 \times 1000 \times 9.81 = 2806 \text{ Pa}$$

Solution (b)
Flow rate

$$Q = u \times a = 0.372 \times [\pi \times (0.3)^2/4] = 0.0263 \text{ m}^3/\text{s}$$
$$= 26.3 \text{ L/s}$$

and mass transfer $M = 26.3$ kg/s.

Solution (c)
From Bernoulli's equation, velocity head $= u^2/2g$ and $u_2 = 9 \times 0.372 = 3.348$ m/s. So gain in velocity head $= (u_2^2 - u_1^2)/2g = [(3.348)^2 - (0.372)^2]/2g = 0.56426$ m of water. And gain in velocity pressure $= h \cdot \rho \cdot g = 0.56426 \times 1000 \times 9.81 = 5535$ Pa. This is equal to the loss in static pressure which is also $dP = 5535$ Pa.
 The manometer is registering the algebraic sum of the shock loss and the loss in static pressure which amounts to $(2806 + 5535)$ Pa or 8.341 kPa. This is the reading of the manometer given in the example and represents a positive displacement of the measuring fluid.

Velocity pressure loss factors

When water flows from a supply pipe into a large tank or when air flows from a supply duct into a room, the shock loss $dh = u^2/2g$ metres of fluid flowing where u is the fluid velocity in the pipe or duct. The fluid velocity beyond

152 Mass transfer of fluids

the pipe or duct is approaching zero. Since the frictional loss is expressed as $dh = k(u^2/2g)$ then in this case

$$(u^2/2g) = k(u^2/2g)$$

and therefore the velocity head loss factor k must equal 1.0. This is known as one velocity head.

In the case of air flow from a duct into a room there will be a grille at the outlet with damper control on the blades. This provides a further loss of energy and the velocity pressure loss factor k will always be in excess of one.

Frictional losses in pipe and duct fittings are expressed as fractions of the velocity head: $(u^2/2g)$ or velocity pressure: $(\rho u^2/2)$. Thus:

$$dP = k \cdot (\rho u^2/2) \text{ Pa} \quad \text{and} \quad dh = k \cdot (u^2/2g) \text{ m of fluid flowing}$$

Since the term k is dimensionless it can be used in either of these formula. It can be defined as the velocity head loss factor or the velocity pressure loss factor.

Example 7.9 Determine the velocity pressure loss factor for a globe valve

The pressure loss across a globe valve located in a horizontal pipe is measured as 126 mbar when flow velocity is 1.9 m/s. Determine the velocity pressure loss factor given water density as 1000 kg/m³.

Solution
Now $dP = k \cdot \rho u^2/2$. Adopting the Bernoulli equation for frictional flow taking sections 1 and 2 as upstream and downstream of the valve, respectively:

$$Z_1 + (P_1/\rho g) + (u_1^2/2g) = Z_2 + (P_2/\rho g) + (u_2^2/2g) + k(u_2^2/2g)$$

Since valve is horizontal $Z_1 = Z_2$, water velocity on either side of the valve is the same; thus, $u_1 = u_2$. Thus the Bernoulli equation reduces to:

$$(P_1 - P_2)/\rho g = k(u^2/2g) \text{ m of water}$$

Note therefore that if there is no change in fluid velocity on either side of the fitting, the frictional loss through it is equal to the static pressure drop. This is not the case for a sudden enlargement or a sudden reduction where a change in velocity does occur on either side of the fitting. See Examples 7.7 and 7.8 in which the fluid velocities u_1 and u_2 in the Bernoulli equation are not the same and therefore do not cancel.
Thus for the globe valve

$$(P_1 - P_2)/\rho g = dh = k(u^2/2g) \text{ m of water}$$

Mass transfer of fluids 153

or $dP = k(\rho u^2/2)$ Pa

substituting:

$$12\,600 = k \times 1000 \times 1.9^2/2$$

from which the velocity pressure loss factor for the globe valve $k = 6.98$.

Example 7.10 Calculate the velocity head loss factor for a gate valve

The displacement of measuring fluid in a differential manometer connected on either side of a 50 mm bore gate valve located horizontally is 2 mm. If the mass transfer of water at 75°C through the valve is 2.87 kg/s, determine its velocity head loss factor. The density of measuring fluid is 13 600 kg/m^3.

Solution

At 75°C water density is 975 kg/m^3 (refer to the *Thermodynamic and Transport Properties of Fluids*). The corresponding displacement of water in the manometer can be determined from $dh = x(S-1)$ metres of water flowing (see Figure 5.4). Thus,

$$dh = 0.002[(13\,600/975) - 1] = 0.0259 \text{ m of water flowing}$$

The volume flow rate of water flowing $Q = 2.87/975 = 0.00294$ m^3/s
  Since $Q = u \times a$, mean velocity

$$u = Q/a = (0.00294 \times 4)/[\pi(0.05)^2] = 1.5 \text{ m/s}$$

Since the fluid velocity on either side of the valve is the same, $u_1 = u_2$, and as the pipe is horizontal $Z_1 = Z_2$, the Bernoulli equation is reduced to:

$$(P_1 - P_2)/\rho g = dh = k \cdot (u^2/2g)$$

and substituting:

$$0.0259 = k \times 1.5^2/2g$$

From which the velocity head loss factor for the gate valve $k = 0.226$.

Example 7.11 Calculate the pressure loss through part of a space heating system

A section of a heating system in which water flows at 0.35 kg/s comprises:
- 1 column radiator, $k = 5$.
- 2–20 mm angle radiator valves, $k = 5$ each.
- 7–20 mm malleable cast-iron bends, $k = 0.7$ each.
- 8–20 mm bore black mild steel pipe in which the coefficient of friction $f = 0.0045$.

Determine the pressure loss due to friction. The water density is given as 975 kg/m³.

Solution
Volume flow of water

$$Q = 0.35/975 = 0.000359 \text{ m}^3/\text{s}$$

Mean water velocity

$$u = Q/a = 0.000359 \times 4/\pi(0.02)^2$$
$$= 1.143 \text{ m/s}$$

The total velocity pressure loss factor for the fittings in the pipe section is found by adding the k values together and $k_t = 19.9$. Assuming no change in fluid velocity in the pipeline and ignoring changes in height, the Bernoulli equation for the fittings is reduced to: $dP = k \cdot \rho u^2/2$ Pa. Then by substitution: the pressure loss attributable to the fittings

$$dP = 19.9 \times 975 \times 1.143^2/2 = 12\,674 \text{ Pa}$$

Similarly, for head loss in straight pipe assuming turbulent flow, the Bernoulli equation is reduced to: $dh = 4fLu^2/2gd$ m of water. Substituting:

$$dh = 4 \times 0.0045 \times 8 \times 1.143^2/2g \times 0.02 = 0.48 \text{ m of water flowing}$$

then

$$dP = dh \cdot \rho \cdot g = 0.48 \times 975 \times 9.81$$

From which the pressure loss in the straight pipe $dP = 4591$ Pa. The total hydraulic pressure loss in the pipe section $dP = 12\,674 + 4591 = 17\,265$ Pa.

Example 7.12 Determine the head loss and gradient as water passes through a pipeline

The mass transfer of water in a 100 mm bore pipe 68 m in length is 30 kg/s. Determine:

(a) The head loss due to friction given water density as 1000 kg/m³, the coefficient of friction f as 0.004 and viscosity μ as 0.00001501 kg/ms. Confirmation should be sought that flow is in the turbulent region.
(b) The gradient to which the pipe must be laid to maintain a constant head.

Mass transfer of fluids 155

Solution (a)

Volume flow $Q = M/\rho = 30/1000 = 0.03$ m³/s

Mean velocity $u = Q/a = 0.03 \times 4/\pi \times 0.1^2 = 3.82$ m/s

Reynolds number $Re = \rho u d/\mu = 1000 \times 3.82 \times 0.1/0/0.00001501$
$= 25\,466\,667$

Since $Re > 3500$, flow is confirmed as being in the turbulent region.

The D'Arcy equation for turbulent flow in straight pipes $dh = 4fLu^2/2gd$ and substituting

$$dh = 4 \times 0.004 \times 68 \times 3.82^2/2 \times 9.81 \times 0.1 = 8.09 \text{ m of water}$$

Head loss due to friction $dh = 8.09$ m of water over a straight pipe of length 68 m.

Solution (b)
Refer to Figure 7.7.

If the pipe is laid to a gradient such that it terminates at a point 8.09 m vertically below the point from which it started, the gradient of $8.09/68 = 0.119$ m/m will ensure that the head loss due to friction is offset by the gradient which is $1/0.119 = 1$ in 8.4. This is called the hydraulic gradient of the pipe.

Example 7.13 Calculate the static regain in a transformation piece carrying air

Determine the regain in static pressure for the transformation piece shown in Figure 7.8 if the velocity pressure loss factor for the fitting is 0.25. The volume flow rate of air at a temperature of 28°C is 3 m³/s and the duct increases gradually in diameter from 0.6 m to 1.0 m. Take air density as 1.2 kg/m³ at 20°C.

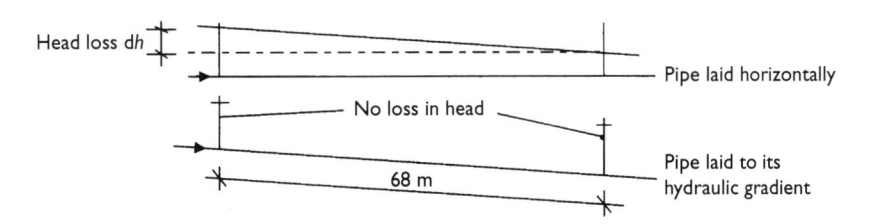

Figure 7.7 The hydraulic gradient (Example 7.12).

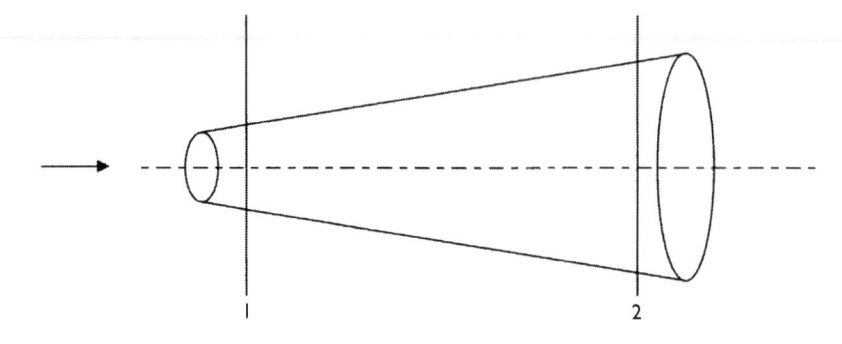

Figure 7.8 Example 7.13.

Solution
For steady flow $Q = u_1 \cdot a_1 = u_2 \cdot a_2$.

$$u_1 = Q/a_1 = 3 \times 4/\pi \times 0.6^2 = 10.61 \text{ m/s}$$
$$u_2 = Q/a_2 = 3 \times 4/\pi \times 1^2 = 3.82 \text{ m/s}$$

Adopting the Bernoulli equation for frictional flow at sections 1 and 2

$$(P_1/\rho g) + (u_1^2/2g) = (P_2/\rho g) + (u_2^2/2g) + k \cdot (u_1^2/2g)$$

Remembering that for an enlargement $P_2 > P_1$ and $u_1 > u_2$. Rearranging

$$(P_2 - P_1)/\rho g = (u_1^2 - u_2^2)/2g - k \cdot (u_1^2/2g)$$

Substituting

$$(P_2 - P_1)/\rho g = [(10.61^2 - 3.82^2)/2g] - (0.25 \times 10.61^2/2g)$$
$$= 4.994 - 1.434$$
$$dh = 3.56 \text{ m of air flowing}$$

Absolute air temperature in the duct $= 273 + 28 = 301$ K. Absolute air temperature at 1.2 kg/m³ and 20°C $= 273 + 20 = 293$ K. Air density correction

$$\rho = 1.2(293/301) = 1.168 \text{ kg/m}^3 \text{ at } 28°C$$
$$dP = 3.56 \times 1.168 \times 9.81 = 41 \text{ Pa}$$

The static pressure regain generated by the transformation piece is 41 Pa. This is distributed along the duct downstream of the fitting.

Taking account of the static loss of air in a rising duct

It will be noticed in Example 7.13 that the terms in the Bernoulli formula for potential energy Z_1 and Z_2 are omitted since the duct is horizontally placed. Clearly they should be accounted for in a rising duct carrying air. Taking a five-storey building having floors at 3 m intervals with the fan (the prime mover) located in the basement and connected to the rising duct, the loss in static pressure in the duct due to the static lift will be $(Z_1 - Z_2) = 6 \times 3 = 18$ m of air or $(dh \cdot \rho \cdot g) = 18 \times 1.2 \times 9.81 = 212$ Pa. If the air in the rising duct is in excess of local ambient temperature, this loss due to the static lift can usually be discounted due to its buoyancy.

Example 7.14 Determine the duty and output power of a booster pump delivering mains water from a ground storage tank

A ground storage tank supplies water to a high level tank in a building, the vertical distance being 25 m. A multistage centrifugal pump is installed on the suction pipe from the low level tank and discharges 2.5 kg/s of water into the 50 mm bore rising main which terminates with a ball valve at the high level tank. Ignore the effects of pressure in the pump suction and determine the net pump duty and output power.

Data: Velocity pressure loss factors – 2 bends $k = 0.4$ each, 2 stop valves $k = 0.7$ each, 1 recoil valve $k = 8.0$.
Assuming turbulent flow, the coefficient of friction in the straight pipe $f = 0.005$. Pressure required at the ball valve $= 30$ kPa. Density of water is 1000 kg/m³.

Solution
The system is shown in Figure 7.9. Adopting the Bernoulli equation for frictional flow and considering sections A and B

$$Z_a + (P_a/\rho g) + (u_a^2/2g) = Z_b + (P_b/\rho g) + (u_b^2/2g) + \text{losses}$$

Now $P_a = P_b =$ atmospheric pressure and the water velocity in the tank u_a approaches zero. Placing all the terms remaining from the Bernoulli equation onto the right-hand side will represent the total energy required for the mass transfer of 2.5 kg/s of water from point A to point B. The net pump head dh that must, therefore, be generated will be:

$$dh = (Z_b - Z_a) + (u_b^2/2g) + \text{losses}$$

The losses include those through the fittings, that through the straight pipe and the discharge pressure required at the ball valve which can be added at the completion of the solution. Thus

$$dh = (Z_b - Z_a) + (u_b^2/2g) + k(u_b^2/2g) + 4fLu_b^2/2gd$$

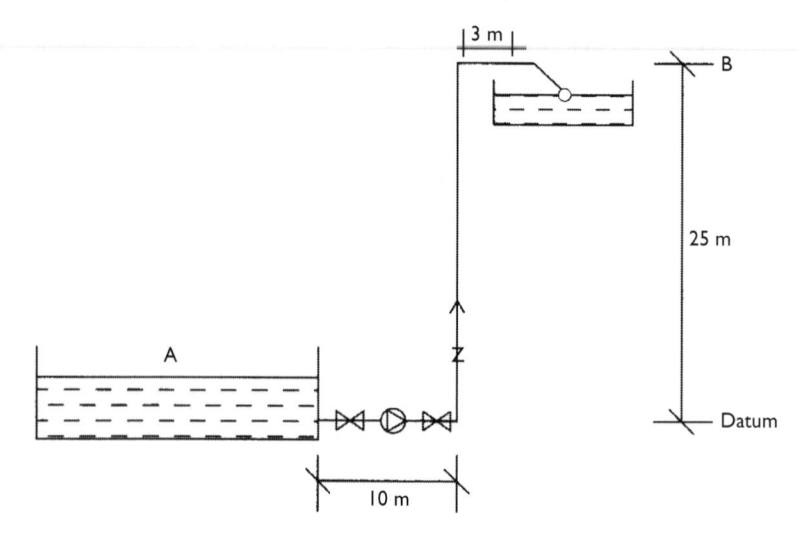

Figure 7.9 Example 7.14.

The total velocity head loss factor is 10.2. You should now confirm that this is so.

Mean velocity of flow

$$u_b = Q/a = 0.0025 \times 4/\pi \times 0.05^2 = 1.273 \text{ m/s}$$

Substituting:

$$dh = (25 - 0) + (1.273^2/2g) + 10.2(1.273^2/2g)$$
$$+ [(4 \times 0.005 \times 38 \times 1.273^2)/(2g \times 0.05)]$$

from which

$$dh = 25 + 0.0826 + 0.8425 + 1.2556$$
$$dh = 27.18 \text{ m of water flowing}$$
$$dP = 27.18 \times 1000 \times 9.81 = 266\,643 \text{ Pa}$$

Net pump pressure = 267 + 30 = 297 kPa and the net pump duty is 2.5 kg/s at 297 kPa. Now from the equation, $dP = dh \cdot \rho \cdot g$, $dh = dP/\rho \cdot g$. Then

$$\text{Pump output power} = dh \cdot M \cdot g$$
$$= (297\,000/1000 \times 9.81) \times 2.5 \times 9.81 = 743 \text{ W}$$

Example 7.15 Sizing the rising main to two high level water storage tanks

Two high level cold water storage tanks having a capacity of 4500 L each are refilled every 2 h. The vertical height of the supply water main will be 26 m and its horizontal distance from the water utilities main is 9 m. If the available pressure is 300 kPa during peak demand, size the rising main.

Data: Assuming turbulent flow, the coefficient of friction $f = 0.007$, allowance for fittings is 30% on the straight pipe, pressure required at the index ball valve is 30 kPa.

Solution
Figure 7.10 shows the system in elevation.
 The volume flow rate

$$Q = (4500 \times 2)/(2 \times 3600) = 1.25 \text{ L/s} = 0.00125 \text{ m}^3/\text{s}$$

Adopting the Bernoulli equation for frictional flow between points A and B

$$Z_a + (P_a/\rho g) + (u_a^2/2g) = Z_b + (P_b/\rho g) + (u_b^2/2g) + \text{loss}$$

Assuming $u_a = u_b$ and rearranging the equation knowing that $P_a > P_b$

$$(Z_a - Z_b) + (P_a - P_b)/\rho g = \text{loss}$$

From equation 6.8, Box's formula for head loss in turbulent flow in straight pipe can be adopted here, thus

$$(Z_a - Z_b) + (P_a - P_b)/\rho g = fLQ^2/3d^5$$

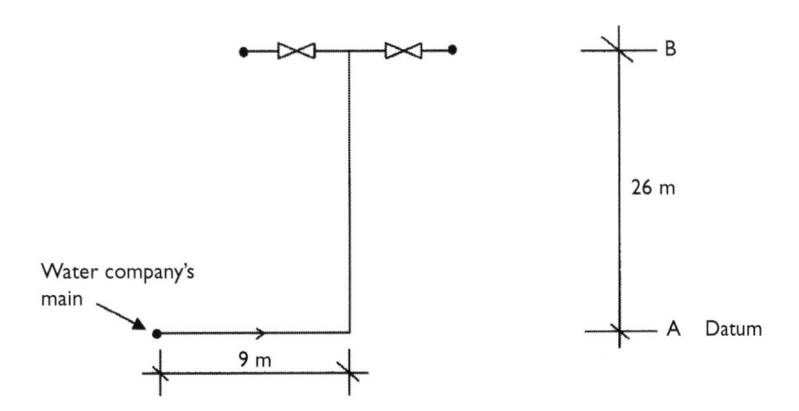

Figure 7.10 Example 7.15.

Total equivalent length of the pipe including fittings $L = (26 + 9) \times 1.3 = 45.5$ m. Substituting:

$$(-26) + [(300\,000 - 30\,000)/1000 \times 9.81]$$
$$= [0.007 \times 45.5 \times (0.00125)^2]/(3 \times d^5)$$

Thus

$$(-26) + 27.52 = 1.6589 \times 10^{-7}/d^5$$

from which

$$d^5 = [1.6589 \times 10^{-7}]/1.52 = 1.0914 \times 10^{-7}$$

and

$$d = 0.0405 \text{ m} = 40.0 \text{ mm}$$

The nearest standard pipe diameter for the rising main = 40 mm.

7.4 Semi-graphical solutions to frictional flow in pipes and ducts

Solutions to problems involving the flow of water in pipes can be undertaken using the Moody chart of Poiseuille and Colebrook–White. This chart is reproduced here from the *Guide* to current practice by kind permission of *CIBSE*. See Figure 6.14.

The calculation routine is as follows: determine the Reynolds number for the known flow conditions from $Re = \rho u d / \mu$, determine the pipe roughness ratio k_s/d, determine the value of the frictional coefficient f from the Moody chart, if the flow is turbulent determine the head loss due to friction from the D'Arcy equation $dh = 4 \cdot f \cdot L \cdot u^2 / 2g \cdot d$. Table 7.1 lists values of absolute roughness k_s, which is a lineal measurement of high points on the rough internal surface, for pipes and ducts.

Example 7.16   Calculate the pressure loss in a horizontal pipe

A horizontal straight pipe 32 mm bore by 2.5 m in length carries water at a mass transfer rate of 1 kg/s. If it is a new black steel pipe, determine the head loss in metres of water, the pressure loss in Pascals and the specific pressure loss in Pa/m.

Data: Absolute viscosity = 0.000378 Ns/m^2, density = 975 kg/m^3.

Table 7.1 Values of absolute roughness in pipes and ducts

Pipes and ducts	k_s (mm)
Copper pipe	0.0015
Plastic pipe	0.003
New black steel pipe	0.046
Rusted black steel pipe	2.5
Clean aluminium ducting	0.05
Clean galvanised ducting	0.15
New galvanised pipe	0.15

In order to use the Moody chart to obtain the coefficient of friction, the relative roughness or roughness ratio and the Reynolds number must be found.

Roughness ratio $= k_s/d = 0.046/32 = 0.00144$

Note that pipe diameter d is left in mm since k_s is measured in mm.
From Chapter 6 the Reynolds number

$$Re = dM/\mu a = [(0.032 \times 1 \times 4)/(0.000378 \times \pi \times 0.032^2)] = 105\,261$$

Given Re and k_s/d, the coefficient of friction f can now be found from the Moody chart, Figure 6.14 from which $f = 0.0055$.

As the flow exceeds the Reynolds number of 3500, flow is clearly turbulent and D'Arcy's equation can be used. Thus $dh = 4fLu^2/2gd$ metres of water flowing, where

$$u = M/\rho a = (1 \times 4)/(975 \times \pi \times 0.032^2) = 1.275 \text{ m/s}$$

and therefore

$$dh = [(4 \times 0.0055 \times 25 \times 1.275^2)/(2 \times 9.81 \times 0.032)]$$
$$= 1.424 \text{ m of water}$$

Pressure loss

$$dP = dh \cdot \rho \cdot g = 1.424 \times 975 \times 9.81 = 13\,6215 \text{ Pa}$$

Specific pressure loss

$$dp = dP/L = 13\,621/25 = 544 \text{ Pa/m}$$

Example 7.17 Determine the specific static pressure loss in a circular duct

Air at a temperature of 8°C flows at 1.02 m³/s along a straight galvanised duct 400 mm in diameter. Determine the specific static pressure loss in Pa/m. Take air density at 20°C as 1.2 kg/m³ and air viscosity at 8°C as 0.00001755 kg/ms.

Solution
Reynolds number $Re = \rho u d/\mu$. The viscosity of dry air which is given here can be obtained from the tables of *Thermodynamic and Transport Properties of Fluids*.

Corrected air density $\rho = 1.2[(273 + 20)/(273 + 8)] = 1.251$ kg/m³

Mean air velocity $u = Q/a = 1.02 \times 4/\pi \times 0.4^2 = 8.12$ m/s

from which $Re = (1.251 \times 8.12 \times 0.4)/(0.00001755) = 232\,186$

Relative roughness $= k_s/d = 0.15/400 = 0.000375$

From the Moody chart, Figure 6.14 the coefficient of friction $f = 0.0044$
Since flow is fully turbulent, D'Arcy's equation can be used and $dh = 4fLu^2/2gd$ where $L = 1.0$ m.
Substituting:

$$dh = (4 \times 0.0044 \times 1 \times 8.12^2)/(2 \times 9.81 \times 0.4)$$

from which for 1 m of straight duct $dh = 0.1479$ m of air flowing and specific pressure loss $dp = dh\rho g/L = (0.1479 \times 1.251 \times 9.81)/1 = 1.815$ Pa/m.

SUMMARY FOR EXAMPLE 7.17
An assumption is made in this solution that the air flowing in the duct is not subject to compression. This is acceptable in ventilation and air conditioning systems but the assumption cannot be made when working with compressed air because of the influence of pressure on air density.

7.5 Gravitational flow in flooded pipes

The mass transfer of water subject to gravity will occur, for example, from a tank located at high level which supplies water to a point at a lower level. It also occurs when a high level reservoir supplies water to a reservoir at some lower level. Without a prime mover such as a pump the static lift provided by the gradient or vertical drop through which the pipe is routed offsets exactly the force of friction opposing fluid mass transfer.

Mass transfer of fluids 163

Example 7.18 Calculate the gravitational mass transfer of water in a connecting pipe

A pipe with a constant gradient connects two reservoirs having a difference in water levels of 20 m. The upper 200 m of pipe is 100 mm bore, the next 100 m is 200 mm bore and the final 100 m is 100 mm bore. The pipes all have coefficients of friction of 0.006. The connections to the reservoirs are bellmouthed and the changes in pipe bore are sudden. Determine the mass transfer of water through the connecting pipe.

Solution

Figure 7.11 shows the system in elevation. Adopting the Bernoulli equation for frictional flow between points A and B

$$Z_a + (P_a/\rho g) + (u_a^2/2g) = Z_b + (P_b/\rho g) + (u_b^2/2g) + \text{losses}$$

$P_a = P_b =$ atmospheric pressure, $u_a = u_b$. And, therefore, these terms will cancel in the Bernoulli equation. Taking Z_b to the left-hand side we have:

$$Z_a - Z_b = \text{losses} = \text{loss in pipe 1} + \text{enlargement loss} + \text{loss in pipe 2}$$
$$+ \text{reduction loss} + \text{loss in pipe 1}$$

Since the exit and entry to the reservoirs are bellmouthed the shock losses due to friction are negligible at these two points. Thus

$$20 = (4fL_1u_1^2/2gd_1) + [(u_1 - u_2)^2/2g] + (4fL_2u_2^2/2gd_2) + (0.5u_1^2/2g)$$
$$+ (4fL_1u_1^2/2gd_1)$$

For steady flow $Q = u_1 \cdot a_1 = u_2 \cdot a_2$, and

$$u_1 = (a_2/a_1) \cdot u_2 = (d_2/d_1)^2 \cdot u_2 = (200/100)^2 \cdot u_2$$

from which $u_1 = 4u_2$

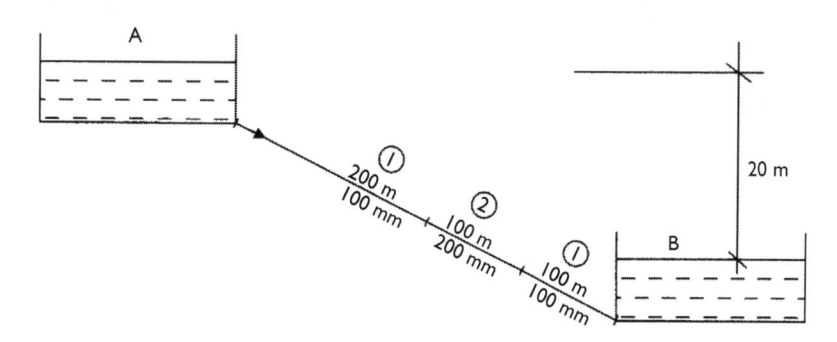

Figure 7.11 Example 7.18.

164 Mass transfer of fluids

Substituting for u and also the other data we have:

$$20 = 39.1u_2^2 + 0.46u_2^2 + 0.61u_2^2 + 0.41u_2^2 + 19.6u_2^2$$

On the right-hand side u_2^2 is a common factor so

$$20 = u_2^2(39.1 + 0.46 + 0.61 + 0.41 + 19.6)$$
$$20 = 60.18u_2^2$$

from which $u_2 = 0.576$ m/s.

$$Q = u \cdot a = 0.576 \times (\pi \times 0.2^2/4) = 0.0181 \text{ m}^3/\text{s}$$

Thus the gravitational flow rate = 18.1 L/s.
Gravitational mass transfer of water = 18.1 kg/s.

Example 7.19 Calculate the total gravitational mass transfer of water in two pipes

Figure 7.12 shows a pipe system connecting two water storage tanks in which the entry and exit losses may be ignored since the connections are bellmouthed. The velocity head loss factor for the bends is 0.3. Determine the total mass transfer of water from the tank at level A to the tank at B.

Data: Pipe 1 is 70 m long and 65 mm bore, f is 0.005. Pipe 2 is 80 m long and 50 mm bore, f is 0.006.

Solution
Adopting the Bernoulli equation for frictional flow, $P_a = P_b =$ atmospheric pressure, and $u_a = u_b$. Thus the available static head will offset exactly the

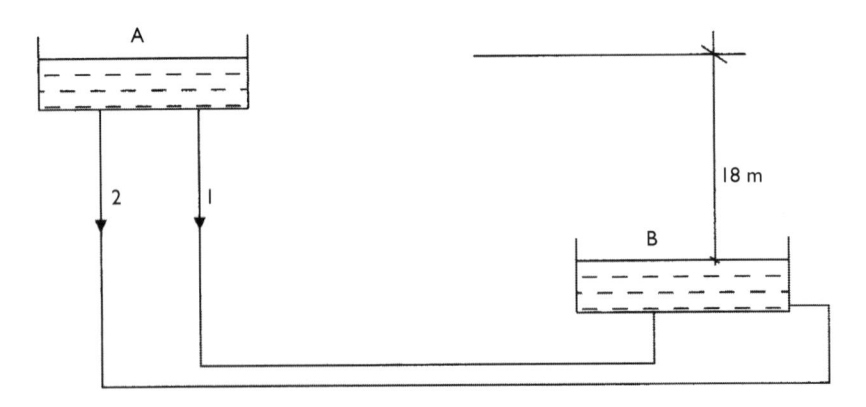

Figure 7.12 Example 7.19.

force of friction opposing the mass transfer of water and therefore:

$$Z_a - Z_b = \text{losses in pipe 1} = \text{losses in pipe 2}$$

since the pipes are in parallel and the frictional losses in each pipe will be similar.

Taking the losses in pipe 1

$$Z_a - Z_b = (4fL_1 u_1^2/2gd_1) + k(u_1^2/2g)$$

where $2g$ is a common factor on the left and may be transposed to the right. Rearranging

$$2g(Z_a - Z_b) = u_1^2(4fL_1/d_1 + k)$$

Rearranging in terms of u_1:

$$u_1 = [2g(Z_a - Z_b)/((4fL_1/d_1) + 2k)]^{0.5}$$

Substituting

$$u_1 = [2g \times 18/((4 \times 0.005 \times 70/0.065) + 0.6)]^{0.5} = 3.99 \text{ m/s}$$

also

$$u_2 = [2g \times 18/((4 \times 0.006 \times 80/0.05) + 0.9)]^{0.5} = 3.0 \text{ m/s}$$

For parallel flow

$$Q = \text{flow in pipe 1 plus flow in pipe 2}$$

$$Q_1 = u_1 \cdot a_1 = 3.99 \times (\pi \times 0.065^2/4) = 0.01324 \text{ m}^3/\text{s}$$

and

$$Q_2 = u_2 \cdot a_2 = 3.0 \times (\pi \times 0.05^2/4) = 0.00589 \text{ m}^3/\text{s}$$

Therefore total gravitational flow

$$Q = Q_1 + Q_2 = 0.01913 \text{ m}^3/\text{s}$$
$$= 19.13 \text{ L/s}.$$

And gravitational mass transfer of water $= 19.13$ kg/s.

Example 7.20 Determine the gravitational mass transfer of water in a vertical pipe

Two tanks, one 15 m vertically above the other as shown in Figure 7.13, are connected together by a 40 mm pipe. Determine the mass transfer of water

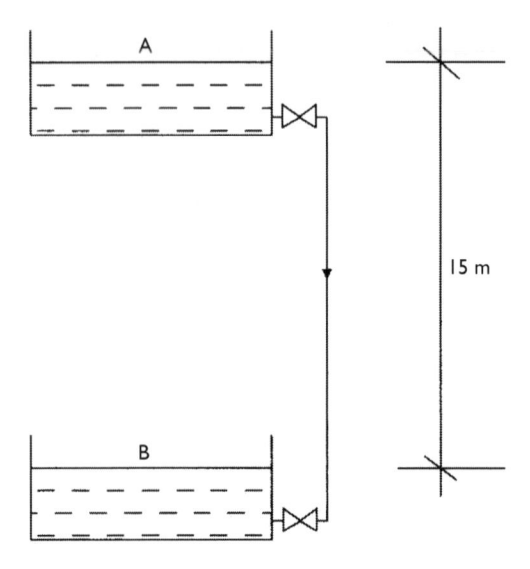

Figure 7.13 Example 7.20.

from the upper to the lower tank. How long will it take to fill the lower tank, given its dimensions are 5 × 2 × 1 m to the water level?

Data: Head loss factors are – for bends 0.3 and for stop valves 5.0.

Solution
Adopting the Bernoulli equation for frictional flow, $P_a = P_b =$ atmospheric pressure, u_a and u_b are approaching zero.
Therefore, rearranging the equation $Z_a - Z_b =$ losses. Thus,

$$Z_a - Z_b = \text{loss in sudden contraction} + \text{loss in pipe} + \text{loss in fittings} + \text{loss in sudden enlargement}$$

and

$$Z_a - Z_b = (0.5u^2/2g) + (4fLu^2/2gd) + (k \cdot u^2/2g) + u^2/2g$$

Note that for the sudden enlargement the water velocity in the tank approaches zero. Thus the shock loss reduces to $u^2/2g$. On the right-hand side $u^2/2g$ is a common factor thus simplifying

$$Z_a - Z_b = u^2/2g \left[0.5 + (4fL/d) + \sum k + 1 \right]$$

Substituting

$$15 = u^2/2g[0.5 + (4 \times 0.005 \times 15/0.04) + 10.6 + 1]$$

from which $u = 3.875$ m/s.
Flow rate will be

$$Q = u \cdot a = 3.875 \times (\pi \times 0.04^2/4) = 0.00487 \text{ m}^3/\text{s}$$

Gravitational flow $Q = 4.87$ L/s; and gravitational mass transfer of water $= 4.87$ kg/s

Time to fill the tank = volume of tank to the waterline/flow rate
$$= 5 \times 2 \times 1/0.00487 = 2053 \text{ s} = 34 \text{ min.}$$

Example 7.21 Determine the gravitational mass transfer of water in two pipes from a common pipeline

Figure 7.14 shows a pipe arrangement connected to a high level tank by which water is discharged to atmosphere at points C and D. Given that the coefficient of friction is 0.005 for all the pipes and that the velocity head loss factor for each bend is 0.7 and that for each valve is 3.0, determine the gravitational mass transfer of water at points C and D.

Data: A–B = 15 m of 25 mm bore pipe, B–C = 21 m of 20 mm bore pipe, B–D = 15 m of 15 mm bore pipe.

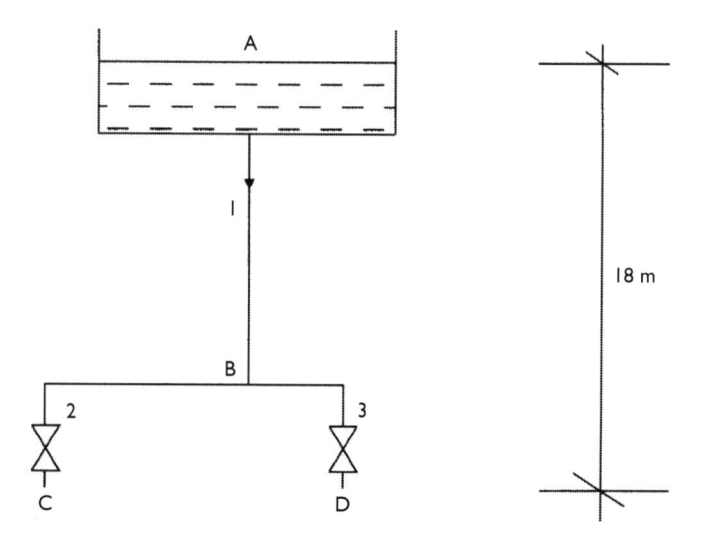

Figure 7.14 Example 7.21.

Solution

From Figure 7.14 pipe sections B–C and B–D are in parallel. Thus $dh_{b-c} = dh_{b-d}$ and $dh_2 = dh_3$. Thus,

$$(4fL_2 u_2^2/2gd_2) + k(u_2^2/2g) = (4fL_3 u_3^2/2gd_3) + k(u_3^2/2g)$$

Simplifying by cancelling $1/2g$

$$(4fL_2 u_2^2/d_2) + ku_2^2 = (4fL_3 u_3^2/d_3) + ku_3^2$$

Identifying the common factors:

$$u_2^2[(4fL_2/d_2) + k] = u_3^2[(4fL_3/d_3) + k]$$

Substituting

$$u_2^2[(4 \times 0.005 \times 21/0.02) + 3.7] = u_3^2[(4 \times 0.005 \times 15/0.015) + 3.7]$$

from which

$$24.7 u_2^2 = 23.7 u_3^2$$

thus

$$u_2^2 = (23.7/24.7) u_3^2$$

and therefore

$$u_2 = 0.9795 u_3$$

For continuity of flow $Q_1 = Q_2 + Q_3$ and $u_1 \cdot a_1 = u_2 \cdot a_2 + u_3 \cdot a_3$, then

$$u_1 \cdot (\pi d_1^2/4) = [u_2 \cdot (\pi d_2^2/4)] + [u_3 \cdot (\pi d_3^2/4)]$$

from which

$$u_1 \cdot d_1^2 = u_2 \cdot d_2^2 + u_3 \cdot d_3^2$$

and

$$u_1 = (1/d_1^2)(u_2 \cdot d_2^2 + u_3 \cdot d_3^2)$$

Substituting for u_2

$$u_1 = (1/d_1^2)(0.9795 u_3 \cdot d_2^2 + u_3 \cdot d_3^2)$$

Substituting for d_1, d_2 and d_3

$$u_1 = 1600(0.000392u_3 + 0.000225u_3)$$

therefore

$$u_1 = 0.9872u_3$$

Now $dh_{a-c} = dh_{a-d}$ and $dh_{1,2} = dh_{1,3}$ since pipes 2 and 3 are in parallel. Working with pipe circuit consisting of sections 1 and 2 and adopting the Bernoulli equation for frictional flow in which $P_a = P_c =$ atmospheric pressure and u_a approaches zero. Note also that the head loss due to water discharge into air at point C is zero. Thus

$$Z_a - Z_c = (u_c^2/2g) + [\text{losses}]$$

The losses include: [sudden contraction + loss in pipe 1 + sudden contraction at tee + loss in pipe 2 + loss in bend and valve]

$$Z_a - Z_c = (u_2^2/2g) + [(0.5u_1^2/2g) + (4fL_1u_1^2/2gd_1) + (0.5u_2^2/2g)$$
$$+ 4fL_2u_2^2/2gd_2 + k \cdot u_2^2/2g]$$

Substituting $u_2 = 0.9795u_3$, $u_1 = 0.9872u_3$, $k = 0.7 + 3 = 3.7$, $f = 0.005$, $L_1 = 15$ m, $L_2 = 21$ m, $d_1 = 0.025$ m, $d_2 = 0.02$ m:

$$18 = 1/2g[0.9795u_3^2 + 0.4873u_3^2 + 11.6948u_3^2 + 0.4797u_3^2$$
$$+ 20.1478u_3^2 + 3.55u_3^2]$$
$$18 = (1/2g) \times 37.3391u_3^2$$

from which

$$u_3 = 3.075 \text{ m/s}$$
$$u_1 = 0.9872u_3 = 0.9872 \times 3.075 = 3.035 \text{ m/s}$$
$$u_2 = 0.9795u_3 = 0.9795 \times 3.075 = 3.012 \text{ m/s}$$
$$Q_c = u_2 \cdot a_2 = 3.012 \times (\pi \times 0.02^2/4) = 0.0009462 \text{ m}^3/\text{s} = 0.95 \text{ L/s}$$
$$Q_d = u_3 \cdot a_3 = 3.075 \times (\pi \times 0.015^2/4) = 0.0005434 \text{ m}^3/\text{s} = 0.54 \text{ L/s}$$

SUMMARY FOR EXAMPLE 7.21
Gravitational mass transfer of water at points C and D are 0.95 kg/s and 0.54 kg/s, respectively.

7.6 Gravitational flow in partially flooded pipes and channels

This section will consider the mass transfer of water in open channels and flooded and partially flooded pipes set to gradients to maintain flow. When there is no prime mover, such as a pump, fluid flow relies on the hydraulic gradient to which the pipe or channel is laid. The gradient offsets exactly the forces of friction opposing flow and generated by the moving fluid.

The Chezy formula is usually associated with this work and this is an adaption of the D'Arcy equation for turbulent flow in pipes where:
$dh = 4fLu^2/2gd$ metres of fluid flowing. The formula can be rearranged in terms of mean flow velocity u thus:

$$u^2 = 2gd \cdot dh/4fL \text{ m/s}$$

It can then be separated into constituent parts:

$$u = (2g/f)^{0.5} \times (dh/L)^{0.5} \times (d/4)^{0.5} \text{ m/s}$$

The term in the D'Arcy equation which can vary in partial flow is $(d/4)$. The constituent parts can then be identified thus: $(2g/f)^{0.5} = C$, the Chezy coefficient. When the coefficient of friction $f = 0.0064$, $C = 55$. You should confirm this value for C.

$dh/L = i =$ the hydraulic gradient m/m

$d/4 = m =$ the hydraulic mean diameter for full and half full bore flow in metres. The hydraulic mean diameter $(d/4)$ will vary according to the volume of partial flow.

The determination of (m) is given below for various flow conditions.

$m =$ (cross-sectional area of flow)/(length of wetted perimeter)

- For full bore flow in conduits of circular cross section $m = (\pi d^2/4)/\pi d = d/4$.
- For half full bore flow in conduits of circular cross section $m = (\pi d^2/8)/(\pi d/2) = d/4$.
- For three quarter full bore flow in conduits of circular cross section it can be shown that $m = d/3$.
- For rectangular channels see (Figure 7.15) hydraulic mean depth $m = d \cdot b/(2d + b)$.

The term m in the Chezy formula must therefore be selected according to the cross-sectional shape of the conduit and the volume of the partial flow. Thus the Chezy formula is given as $u = C \cdot (m \cdot i)^{0.5}$ m/s. Invariably, the Chezy coefficient is taken as $C = 55$.

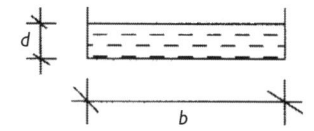

Figure 7.15 Hydraulic mean depth in rectangular channels $m = db/(2d + b)$.

Example 7.22 Calculate the discharge capacity of a drain pipe set to a gradient

Determine the discharge capacity of a 160 mm drain flowing half full bore when it is laid to a gradient of 1 : 150. Take the Chezy coefficient C as 55.

Solution
The drain may be taking rainwater run-off or it may be connected to a vertical soil stack.

In either case the gradient is required to ensure flow to the point of discharge. From the Chezy formula $u = 55 \cdot (m \cdot i)^{0.5}$. Substituting, mean flow velocity

$$u = 55[(0.16/4)(1/150)]^{0.5} = 55 \times 0.01633 = 0.898 \text{ m/s}$$

For half full bore, flow in the drain

$$Q = u \cdot a = 0.898 \times (\pi(0.16)^2/8) = 0.009 \text{ m}^3/\text{s}$$
$$= 9 \text{ L/s}$$

Gravitational mass transfer of water = 9 kg/s.

Example 7.23 Determine the required gradient and mass transfer of soil water in a drain pipe

(a) Determine the gradient required for a 110 mm drain to run a three quarter full bore at a mean soil water velocity of 1.2 m/s. The coefficient of friction f for the drain pipe is 0.008.
(b) Determine the mass transfer of soil water.

Solution (a)
The Chezy coefficient $C = (2g/f)^{0.5} = (2g/0.008)^{0.5} = 49.523$. From the Chezy formula

$$u^2 = C^2 \times m \times i$$

Rearranging: $i = u^2/C^2 \cdot m$ where for three quarter full bore $m = d/3$.

Substituting: $i = 1.2^2 \times 3/(49.523)^2 \times 0.11 = 0.016013$ m/m

172 Mass transfer of fluids

Now the hydraulic gradient $i = dh/L = 0.016013$. Therefore for a fall of 1 m, drain length $L = dh/0.016013 = 1/0.016013 = 62.4$. Thus the minimum hydraulic gradient for the drain is 1 : 62.4.

Alternative solution to part (a). The solution to this problem can also be done by adopting the Bernoulli equation for frictional flow in which the head loss sustained can be obtained from the D'Arcy equation. Consider two points along the drain:

$$Z_a + (P_a/\rho g) + (u_a^2/2g) = Z_b + (P_b/\rho g) + (u_b^2/2g) + \text{loss}$$

Now $P_a = P_b$ and $u_a = u_b$. Thus rearranging the Bernoulli equation and cancelling P and u terms: $Z_a - Z_b = \text{loss}$.

The D'Arcy equation for turbulent flow is $dh = 4fLu^2/2gd$ metres of water flowing, but the hydraulic mean radius for three quarter full bore is $m = d/3$. Therefore $dh = 3fLu^2/2gd$ and considering 1 m length of drain pipe and substituting:

$$Z_a - Z_b = (3 \times 0.008 \times 1.0 \times 1.2^2)/(2g \times 0.11)$$

and the vertical fall $Z_a - Z_b = 0.016013$ m.

For a 1 m vertical fall, the length of the drain pipe $L = 1/0.016013 = 62.4$ m. Thus the minimum gradient to which the pipe must be laid to achieve a mean velocity of 1.2 m/s will be 1 : 62.4.

Solution (b)
Flow rate for three quarter full bore

$$Q = u \cdot a = 1.2 \times 0.75(\pi \times 0.11^2/4) = 0.00855 \text{ m}^3/\text{s}$$

Gravitational mass transfer of soil water = 8.55 kg/s. Flow in vertical soil stacks is subject to the full impact of gravitational acceleration.

In such circumstances, gravitational acceleration and opposing friction will balance and a terminal velocity is reached.

For soil stacks which contain air and are open to atmosphere, in order to reduce water and air disturbance, stack loading is taken as about one quarter full. Due to the coanda effect the water slides down the inside of the stack wall leaving a core of air in the centre. See Figure 7.16.

The empirical formula for sizing the vertical soil stack when flowing one quarter full is $Q = K \cdot d^{(8/3)}$ L/s. The constant $K = 0.000032$ for quarter full flow and $K = 0.000052$ for one-third full flow. Soil stack diameter d is in mm.

Example 7.24 Calculate the size of a vertical stack, given simultaneous flow

The simultaneous discharge of soil water into a vertical stack is estimated as 6.5 L/s. Determine the size of the stack to accomodate the discharge.

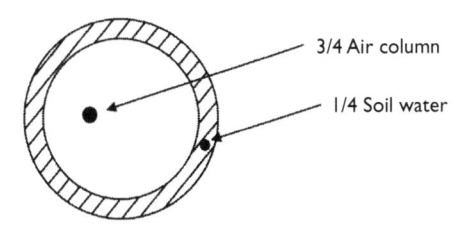

Figure 7.16 Vertical soil stack – accommodation of air and water.

Table 7.2 Impermeability of different surfaces

Type of surface	Average P
Watertight	0.9
Asphalt	0.875
Closely joined stone	0.825
Macadam roads	0.435
Lawns	0.15
Wood	0.105

Solution

From the empirical formula for flow in vertical soil stacks

$$d = (Q/K)^{(3/8)} = (6.5/0.000032)^{(3/8)} = 97.8 \text{ mm}$$

The nearest standard stack diameter $d = 100$ mm.

Rainwater run-off depends upon the surface on which it lands. The amount of water which can be expected from any given surface depends upon

- the area of surface upon which rain is falling;
- the surface type;
- whether the surface is level or sloping;
- the rainfall intensity and
- the rate of evaporation, which is seasonal.

Rate of run-off $Q = A \cdot P \cdot R / 3600 \times 1000 \text{ m}^3/\text{s}$

The impermeability P of the surface depends upon its type. Refer to Table 7.2.

Example 7.25 Size the rain water drain pipes serving a car park

A car park having an asphalt surface measures 50 m × 30 m and is laid to ensure adequate water run-off into a drainage channel running continuously

174 Mass transfer of fluids

along the length of the parking area. Determine the gradient and diameter of each of the main drain pipes connected to each end of the drainage channel assuming the high point is half way along its length.

Data: rainfall intensity is to be taken as 75 mm/h, mean water velocity in the drain pipes is 1.1 m/s and the drains are to run three quarter full bore; the Chezy coefficient $C = 55$.

Solution
Rainfall run-off

$$Q = A \cdot P \cdot R/3600 \times 1000 = (50 \times 30) \times 0.875 \times 75/3600 \times 1000$$
$$= 0.02734 \text{ m}^3/\text{s}$$

Each drain must handle $0.02734/2 = 0.01367$ m^3/s and for each drain $Q = u \cdot a$ where the wetted cross section of the drain is 3/4.

Substitute $0.01367 = 1.1 \times 0.75(\pi d^2/4)$

from which $d = [(0.01367 \times 4)/(1.1 \times 0.75 \times \pi)]^{0.5}$

and $d = 0.145$ m $= 145$ mm

If the nearest standard size of drain pipe is 160 mm, the mean water velocity will be:

$$u = Q/a = (0.01367 \times 4)/(0.75\pi \times 0.16^2)$$

and mean velocity $u = 0.906$ m/s. Finding the gradient to which each drain pipe must be laid can be done by adopting the Chezy formula

$$u = C \cdot (m \cdot i)^{0.5}$$

from which

$$i = u^2/(C^2 \cdot m) \quad \text{m/m}$$

where for three quarter flow $m = d/3$.
Therefore

$$dh/L = (0.906^2 \times 3)/(55^2 \times 0.16) = 0.0050878$$

from which

$$L = 1/0.0050878 = 197$$

and the hydraulic gradient is 1 : 197.

SUMMARY TO EXAMPLE 7.25
The gradient of 1:197 is the minimum gradient for the drain to achieve a mean water velocity of 0.906 m/s. If the gradient was increased to 1:100 the mean water velocity increases to 1.27 m/s.

You should now confirm this calculation.

The rainfall intensity of 75 mm/h is not to be considered as acceptable in all design solutions. Rainfall intensities equivalent to 250 mm/h are possible in the United Kingdom although they may last for only a few moments or even seconds. The actual figure selected from Met. data will depend upon from where the run-off is collected.

For example, a single storey flat roofed building against a gable end consisting of two or more storeys will pick up rain water run-off from the gable wall when the wind is driving rain against it. This increases the run-off from the flat roof.

Example 7.26 Determine the mass transfer of water in an open channel

An open channel as shown in Figure 7.17 is laid to a gradient of 1:80. If the maximum depth of water flowing is to be 80 mm, determine the mass transfer of water in the channel. Take the Chezy coefficient as 55.

Solution
The mean hydraulic depth $m =$ (cross-sectional area of flow)/(length of wetted perimeter). From Figure 7.17, $x = 80 \tan 30 = 80 \times 0.5774 = 46$ mm, thus cross-sectional area of flow $= (150 + 46) \times 80 = 15\,680$ mm^2 $= 0.0157$ m^2. Also from Figure 7.17, $y = 80/\cos 30 = 92$ mm, thus the wetted perimeter $= 150 + 92 + 92 = 334$ mm $= 0.334$ m. The hydraulic mean depth, $m = 0.0157/0.334 = 0.047$.

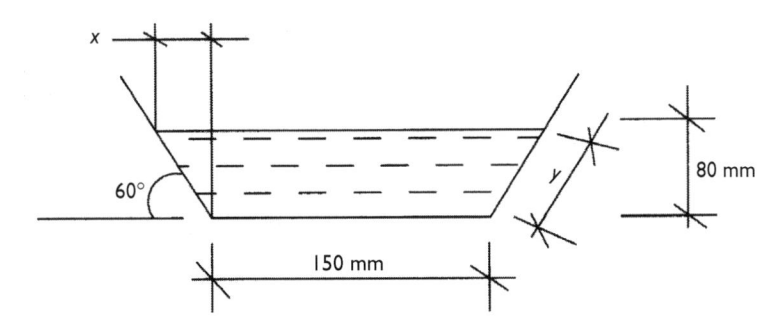

Figure 7.17 Example 7.26.

176 Mass transfer of fluids

Adopting the Chezy formula for turbulent flow in open channels:

$$u = C(m \cdot i)^{0.5}$$

Substituting $u = 55(0.047 \times 1/80)^{0.5} = 1.333$ m/s

Depth of water in the channel is 80 mm and width $= (150 + 0.46) = 196$ mm. Flow rate

$$Q = u \cdot a = 1.333 \times (0.08 \times 0.196) = 0.0209 \text{ m}^3/\text{s}$$

Taking the density of rain water as 1000 kg/m^3, the mass transfer of water in the channel $= 21$ kg/s.

7.7 Alternative rational formulae for partial flow

There are three formulae which are adaptions to the Chezy formula and the Darcy equation.

1. The Manning formula, where

    $$u = (1/n)(m^{0.667})(i^{0.5}) \text{ m/s}$$

    where $n =$ the roughness coefficient

    $\quad\quad = 0.009$ for glass smooth pipe

    $\quad\quad = 0.022$ for dirty cast-iron pipe.

    The roughness coefficient n is not to be confused with the coefficient of friction f. The Chezy coefficient $C = (2g/f)^{0.5} = 1/n$ and when $C = 55$, $n = 1/55 = 0.0182$. Also $n = (f/2g)^{0.5}$, and therefore $f = 2g \cdot n^2$. You should now confirm these relationships.

    These relationships between n, f and C are tabulated for four values of C and are shown in Table 7.3. You should confirm the relationships in Table 7.3.

    Table 7.3 Relationships between the coefficients

n	f	C
0.009	0.0016	111
0.022	0.0095	45.5
0.0182	0.0065	55
0.012	0.0028	84

2 The Crimp and Bruge's formula took the roughness coefficient n as 0.012 and adapted the Chezy equation thus:

$$u = 84(m^{0.667})(i^{0.5}) \text{ m/s}$$

3 The D'Arcy–Weisback formula adopts both the D'Arcy equation for turbulent flow and the Chezy equation, thus:

$$u = (2g \cdot m \cdot i/f)^{0.5} \text{ m/s}$$

This formula is in fact the same as the D'Arcy equation from which the Chezy formula is derived, thus $u = [(2g(d/4)(dh/L)(1/f)]^{0.5}$ m/s, from which the D'Arcy equation $dh = 4fLu^2/2gd$ metres of fluid flowing is obtained. You should now confirm that the D'Arcy–Weisback formula is the same as the D'Arcy equation.

You should also note that the D'Arcy equation and the D'Arcy–Weisback formula are given with the hydraulic mean diameter $m = d/4$. This value for m only applies to full bore and half bore flow in circular conduits. A flow of three quarter bore in a circular conduit, for example, has $m = d/3$ and flow in a channel of rectangular cross section has $m = db/(2d + b)$.

These substitutions must be made before adopting either of these formulae for partial flow.

Example 7.27 Compare the mean fluid velocity in a 100 mm bore pipe set to a gradient using four rational formulae

Determine the mean fluid velocity of flow in flooded pipes and pipes carrying fluid at half full bore by adopting the following formulae and compare the results.
Take the Chezy coefficient as 55 and the hydraulic gradient as 1 : 50.
Chezy formula, Manning formula, Crimp and Bruges formula, D'Arcy–Weisback formula.

Solution
The results are tabulated and given in Table 7.4
You should now confirm the solutions given in Table 7.4. Remember that for flooded pipes and pipes carrying fluid at half full bore $m = d/4$. Note the similarity of solution between Chezy and D'Arcy–Weisback.

Example 7.28 Compare the mean water velocity and mass transfer in a channel set to a gradient using four rational formulae

Determine and compare the mean water velocity and rate of flow in a rectangular channel 150 mm wide and having a water depth of 50 mm when the hydraulic gradient is 1 : 100. Take the Chezy coefficient

178 Mass transfer of fluids

Table 7.4 Comparison of fluid velocity, Example 7.27

Source	Mean fluid velocity (m/s)
Chezy	1.23
Manning	0.664
Crimp and Bruges	1.014
D'Arcy–Weisback	1.23

as 45.5. The comparison should be taken using the rational formulae for partial flow.

Solution
The hydraulic mean depth for rectangular channels

$$m = db/(2d+b) = (0.05 \times 0.15)/[(2 \times 0.05) + 0.15] = 0.03$$

Adopting the Chezy formula

$$u = C(m \cdot i)^{0.5}$$

substituting

$$u = 45.5[0.03 \times (1/100)]^{0.5} = 0.788 \text{ m/s}$$

partial flow

$$Q = u \cdot a = 0.788 \times 0.05 \times 0.15 = 5.91 \text{ L/s}$$

Adopting the Manning formula

$$u = m^{0.667} \times i^{0.5} \times 1/n$$

now

$$C = (2g/f)^{0.5}$$

from which

$$f = (2g/C^2) = 2 \times 9.81/45.5^2 = 0.0095$$

and

$$n = (f/2g)^{0.5} = (0.0095/2 \times 9.81)^{0.5} = 0.022$$

substituting

$$u = (0.03^{0.667}) \times (1/100)^{0.5} \times (1/0.022)$$

from which

$$u = 0.438 \text{ m/s}$$

partial flow

$$Q = u \cdot a = 0.438 \times (0.05 \times 0.15) = 3.29 \text{ L/s}$$

Adopting the Crimp and Bruge's formula

$$u = 84(m)^{0.667} \times (i)^{0.5}$$

substituting

$$u = 84 \times 0.03^{0.667} \times (1/100)^{0.5}$$

from which

$$u = 0.81 \text{ m/s}$$

partial flow

$$Q = u \cdot a = 0.81 \times (0.05 \times 0.15) = 6.08 \text{ L/s}$$

Adopting the Darcy–Weisback formula

$$u = (2g \cdot m \cdot i/f)^{0.5}$$

substituting

$$u = [2 \times 9.81 \times 0.03 \times (1/100)(1/0.0095)]^{0.5}$$

from which

$$u = 0.787 \text{ m/s}$$

partial flow

$$Q = u \cdot a = 0.787 \times (0.05 \times 0.15) = 5.9 \text{ L/s}$$

Summarising solution to Example 7.28 in Table 7.5.
Note the discrepancy between the Manning formula and the others in Table 7.5.

Table 7.5 Comparison of solutions for Example 7.28

Rational formula	Mean velocity (m/s)	Partial flow (kg/s)
Chezy	0.788	5.91
Manning	0.438	3.29
Crimp and Bruge's	0.81	6.08
D'Arcy–Weisback	0.787	5.9

7.8 Flow of natural gas in pipes

The chemical composition of natural gas varies only slightly along with its calorific value of 39 MJ/m^3. Natural gas density is approximately 0.7 kg/m^3 at 20°C and standard atmospheric pressure. The combustion of fossil fuels and the consequent carbon dioxide emissions is covered in another book in the series. This section focuses on the flow of gas in pipes. There are two formulae associated with gas flow, which is in the turbulent region, namely

- Box's formula 6.8 where $Q = (3hd^5/fL)^{0.5}$ m^3/s and
- Pole's formula which is identified in *CIBSE Guide G* as

$$Q = 0.001978d^2(Hd/SL)^{0.5} \text{ L/s}$$

The term S in Pole's formula is the ratio of relative densities of gas and atmospheric air and $S = 0.7/1.2 = 0.58$ at an ambient temperature of 20°C. Both Box's and Pole's equations can be reconciled although it is a drawn out process. You might like to try a reconciliation. Box's and Pole's formulae assume that the fluid flowing is incompressible. At low pressure the mass transfer of both natural gas and atmospheric air in pipes and ducts is assumed to be under incompressible conditions.

Natural gas on the service side of the meter is normally about 21 mbar (2100 Pa) gauge with an allowable pressure drop between the meter and the appliance of 1 mbar (100 Pa). For gas mains operating at higher pressures recourse is made to the sizing procedure for compressed air. See Section 7.9.

The total allowable pressure drop in the system on the service side of the meter may be increased if the gas main rises significantly within the building. The gain in pressure

$$P = (\rho_{air} - \rho_{gas})g \text{ Pa/m} = (1.2 - 0.7) \times 9.81 = 4.9 \text{ Pa/m}.$$

So for a roof top plant room in a 10-storey building allowing 3 m per floor the gas pressure rise is 147 Pa making a total of (100 + 147) = 247 Pa available for pipe sizing the gas main in the plant room.

Example A2.8 in Appendix 2 analyses the development of Pole's formula which will be used to size pipes carrying natural gas.

Mass transfer of fluids 181

The remaining terms in the formula above are: $d =$ pipe diameter in mm; $L =$ pipe length in m; $H =$ pressure loss in mbar.

There now follows an example that compares gas pipe sizing from the *CIBSE Guide* section C4 with the use of Pole's formula.

Example 7.29 Verifying Pole's formula for sizing a pipe carrying natural gas

Pipe diameter is 32 mm from the CIBSE pipe sizing tables for medium grade black tube transporting natural gas when $Q = 0.00527$ m^3/s and specific pressure drop dp is 8 Pa/m. Verify the pipe diameter using Pole's formula.

Data: $L = 1$ m, $S = 0.58$.

Solution

Since the subject of Pole's formula is Q L/s it needs to be rearranged with diameter d as the subject.

$$Q = 0.001978 d^2 (Hd/SL)^{0.5}$$
$$Q^2 = 0.0000039 d^4 (Hd/SL)$$
$$Q^2 = 0.0000039 d^5 (H/SL)$$
$$d^5 = 255\,592 Q^2 (SL/H)$$
$$d = 12.0645 (Q)^{2/5} (SL/H)^{1/5}$$

Substituting $d = 12.0645 (5.27)^{2/5} (SL/H)^{1/5}$, where $Q = 5.27$ L/s. From which

$$d = 12.0645 \times 1.944 \times (0.58 \times 1/(0.08))^{1/5}$$

where $H = 8$ Pa $= 0.08$ mbar.

Thus $d = 12.0645 \times 1.944 \times 1.48616$ and finally $d = 34.86$ mm.

Clearly this is slightly larger than the pipe size extrapolated from the pipe sizing tables. However, it is closer to 32 mm than the next standard pipe size up which is 40 mm.

7.9 Flow of compressed air in pipes

The mass transfer of compressed air in pipes must account for its compressibility. To allow for this, pressure factor $Z = P^{1.929}$ and pressure factor loss per unit length of pipe is dZ/L or $(Z_1 - Z_2)/L$.

Note: This is different to the flow of fluids considered to be incompressible in which pressure loss per unit length of pipe is expressed as dP/L Pa/m.

Air flow through a pipework system varies with temperature and pressure. However, temperature variations are negligible and so isothermal conditions prevail and Boyle's law applies; thus $P_1V_1 = P_2V_2$ where V = volume and can be substituted by Q m^3/s. So $Q_2 = Q_1(P_1/P_2)$ where Q_1 is the free air being compressed, Q_2 is the compressed air, P_1 is atmospheric pressure (101 kPa) and P_2 is the absolute pressure of the compressed air in the pipe.

From the generic formula $Q = u \cdot a$ m^3/s, and therefore $Q_1 = u \cdot (\pi d^2/4)$

So rearranging $u = Q_1 \times (4/\pi d^2)$ m/s
Substituting for Q_2: $u = Q_1(P_1/P_2) \times (4/\pi d^2)$ m/s

where u = mean air velocity (limiting value is 6 m/s).
If pipe diameter is required then by rearranging the equation for velocity and taking the square root:

$$d = [Q_1(P_1/P_2)(4/\pi u)]^{0.5} \text{ m}$$

Example 7.30 Sizing a compressed air pipeline

A compressed air line serves equipment that operates at 2.3 bar gauge and is required to convey 17 L/s of free air.

(a) Determine a suitable pipe size for a limiting velocity of 6 m/s.
(b) Calculate the specific pressure drop over a total equivalent length of 20 m.
(c) Find the actual mean velocity of the compressed air in the selected pipe.

Solution (a)
From the equation for pipe diameter $d = [0.017(101/330)(4/6\pi)]^{0.5}$
From which diameter $d = 0.33$ m or 33 mm
The nearest standard pipe size is 32 mm.

Solution (b)
From Table 7.6 for 32 mm pipe conveying 0.017 m^3/s of air $dZ/L = 22.5$.
Pressure required at the equipment P_2 is 2.3 bar gauge which is 330 kPa abs.

$$P_2 = 330 \text{ kPa}, \; Z_2 = P_2^{1.929} = 72\,146;$$

so

$$Z_1 = 72\,146 + (22.5 \times 20) = 72\,596$$

$$P_1 = {}^{1.929}\!\sqrt{Z_1} = {}^{1.929}\!\sqrt{72\,596} = 331.065 \text{ kPa}$$

Specific pressure loss

$$dp = (331.065 - 330)/20 = 0.05325 \text{ kPa/m} = 53 \text{ Pa/m}$$

Mass transfer of fluids 183

Table 7.6 Extract from *CIBSE Guide* section C4 flow of compressed air in black heavy weight tube

dZ/L	Volume flow in m^3/s for the following pipe diameters (mm)				
	15	20	25	32	40
10	0.001	0.003	0.005	0.011	0.017
20	0.002	0.004	0.008	0.016	0.024
25	0.002	0.004	0.009	0.018	0.027

Solution (c)
From above, mean air velocity

$$u = Q_1(P_1/P_2)(4/\pi d^2)$$

Substituting

$$u = 0.017(101/330)(4/(0.032)^2\pi) = 6.47 \text{ m/s}$$

Note: If 32 mm pipe is used, the mean air velocity exceeds the limiting velocity of 6 m/s. The next standard pipe size up is 40 mm. You should now revisit part (b) and part (c) using 40 mm diameter pipe. The answers are (b) = 23.67 Pa/m and (c) 4.14 m/s.

7.10 Vacuum pipe sizing

Flow through a pipework system varies with temperature and pressure. However, temperature variations are negligible and so isothermal conditions prevail as they do with compressed air, and Boyle's law applies. Thus $P_1V_1 = P_2V_2$. So $Q_2 = Q_1(P_1/P_2)$ where Q_1 is the free air, Q_2 is the evacuated air in m^3/s, P_1 is atmospheric pressure and P_2 is the pressure in the vacuum pipe. Taking atmospheric pressure as 101 kPa or 760 mm mercury, the design of vacuum lines uses the term Torr where the classification of vacuum is

low or rough vacuum: 90–125 Torr
medium vacuum: 10–15 Torr
high vacuum: up to 1 Torr

and 760 Torr represents atmospheric pressure with 0 Torr being ultimate vacuum.

Since $Q = u \cdot a = u(\pi d^2/4)$ m^3/s, by rearranging, pipe diameter $d = [(Q_2)(4/\pi u)]^{0.5}$ m.

By substitution $d = [Q_1(P_1/P_2)(4/\pi u)]^{0.5}$ m and by rearranging $u = Q_1(P_1/P_2)(4/\pi d^2)$ m/s, where the limiting velocity is 25 m/s.

These formulae are similar to those used for compressed air.

Example 7.31   Sizing a pipeline under vacuum

An operating theatre has two vacuum points each rated at 1.34 L/s of free air and working at 125 Torr. Size the common pipe on an air velocity of 20 m/s.

Solution

The volume flow rate of free air $Q_1 = 2 \times 1.34 = 2.68$ L/s $= 0.00268$ m^3/s. From the equation above for pipe diameter: pipe diameter $d = [0.00268 \times (760/125)(4/20\pi)]^{0.5} = 0.0322$ m.

So the standard pipe diameter will be 32 mm.

7.11   Chapter closure

You have been introduced to the mass transfer of fluids, both compressible, non-compressible and vacuum in pipes and ducts and to the partial mass transfer of water in pipes, soil stacks and open channels.

This chapter has introduced the subject matter from first principles. In practice, of course, it is recognised that, for convenience and to save time, the use of tables and software is required for design work. However, this chapter investigates the underpinning theory of the mass transfer of fluids.

Successful conclusion of this chapter will enable you to tackle a number of practical problems from first principles for which there may be no alternative but to undertake manual solutions.

Chapter 8

Natural ventilation in buildings

Nomenclature

A	area (m²)
A_1, A_2, A_3, A_4	free area of openings (m²)
A_s	free area subject to stack effect (m²)
A_w	free area subject to wind (m²)
C	specific heat capacity (kJ/kgK)
C_d	0.61, coefficient of discharge
dC_p	difference in pressure coefficient
dP	difference in pressure (Pa)
dP_u	difference in velocity pressure (Pa)
dt	difference in temperature (K)
$d\rho$	difference in air density (kg/m³)
F	force in Newtons (N)
H	heat energy (kWh, kJ)
h	height (m)
J	factor for degree of openable window
LMTD	log mean temperature difference (K)
M	mass transfer (kg/s)
N	air change rate per hour (h⁻¹)
P	pressure (Pa)
P_i	pressure of air column indoors (Pa)
P_o	pressure of equivalent air column outdoors (Pa)
P_u	velocity pressure (Pa)
Q_p	plant energy output (kW)
Q_s	volume flow of air due to the stack effect (m³/s)
Q_w	volume flow of air due to wind m³/s
t_i	customary indoor temperature (°C)
T_i, T_o	absolute indoor/outdoor temperature (K)
t_m	mean temperature (°C)
t_o	customary outdoor temperature (°C)
u	air velocity (m/s)

u_h	air velocity at roof level (m/s)
u_m	mean air velocity (m/s)
V	volume (m^3)
ρ	air density (kg/m^3)
ρ_i	air density indoors (kg/m^3)
ρ_o	air density outdoors (kg/m^3)

8.1 Introduction

With the improvement in standards of thermal insulation for the building envelope, the proportion of the plant energy output Q_p required to offset heat loss resulting from natural infiltration of outdoor air has increased. Modern buildings are better sealed against the random infiltration of outdoor air, but the trend towards the increasing proportion of the building heat loss which has to account for natural infiltration is continuing, since adequate ventilation of the building shell is essential, whereas there is no limitation upon the improvement in the thermal insulation standards of the building envelope.

There are a number of factors which influence the rate of natural ventilation in a building:

- Wind speed and direction, influenced by geographical location, with respect to the orientation of the building.
- The buoyancy forces or stack effect which induces natural draught within the building and depends upon the difference between indoor and outdoor temperature.
- The height of the building.
- The shape and location of the building with respect to buildings in the vicinity.
- Wind breaks, natural and artificial.
- How well the building is sealed.

The design of the lift shaft, stair wells and atrium particularly in tall buildings, can have a significant effect upon infiltration initiated by the wind and/or the stack effect.

8.2 Aerodynamics around a building

As air flows over and around a building, it creates positive and negative zones of pressure. Figure 8.1 shows typical wind and pressure patterns in both elevation and plan. Figure 8.2 shows how the prevailing air flow divides over and around a building with the location of the line of maximum air pressure on the facade facing the windward side.

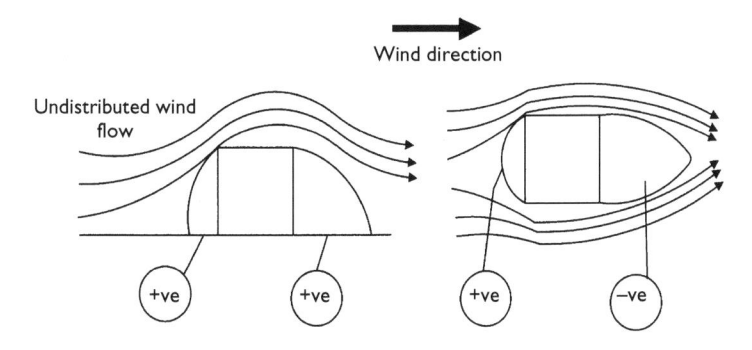

Figure 8.1 Air movement round a building producing +ve and −ve pressures.
Source: Reproduced with permission of the Heating and Ventilating Contractors' Association.

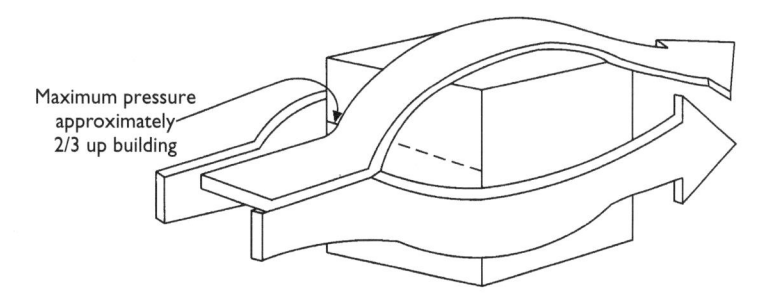

Figure 8.2 Typical air movement over and round a building, one-third over roof, two-thirds round sides.
Source: Reproduced with permission of the Heating and Ventilating Contractors' Association.

Positive pressure (+ve) is created on the windward face and air flow separation occurs at the corners, eaves and roof ridge. Negative pressures (−ve) are generated by air separation along the sides of the building, over the ridge and on the leeward face. Refer to a building in plan and elevation subject to the effects of wind in Figure 8.3.

Figure 8.4 shows the pressure effects of wind when it is incident on the corner of a building. The effect of wind pressure on a building will have a significant baring upon the natural ventilation occurrence. Zones of negative pressure can cause pollution in some rooms on the leeward side from exit points of mechanical extract and from the products of combustion emanating from chimneys. Figure 8.5 shows a building in elevation and the potential zone subject to the effects of pollution from the point of extract. The wind velocity profile shown in Figure 8.5 will vary with the roughness of the underlying surfaces, or terrain.

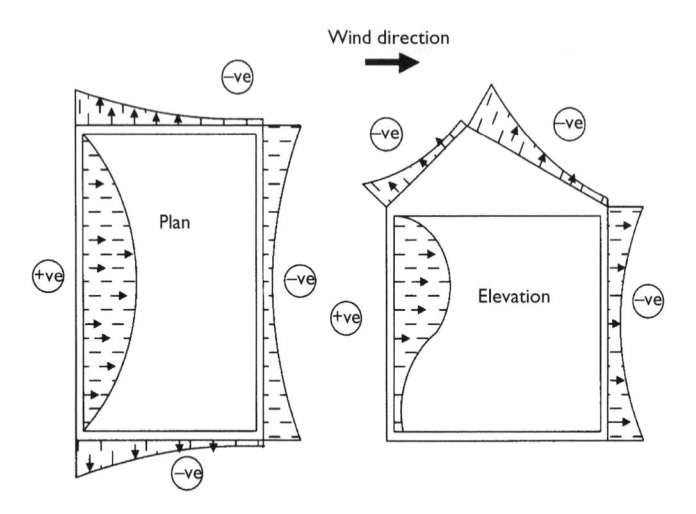

Figure 8.3 Wind pressure distribution on a building.

Source: Reproduced with permission of the Heating and Ventilating Contractors' Association.

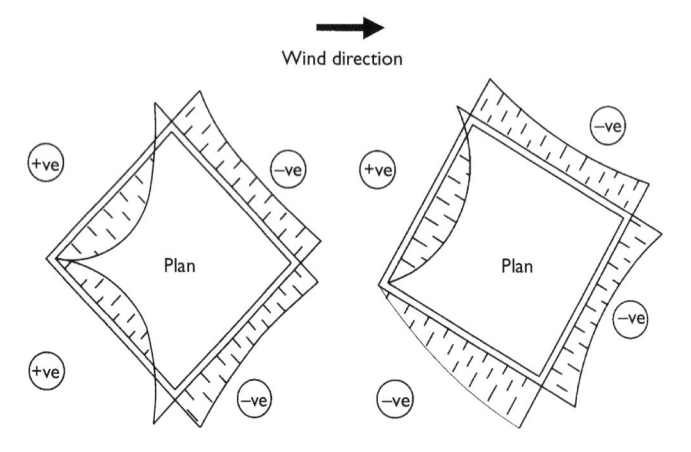

Figure 8.4 Pressure distribution with wind on corner of building.

Source: Reproduced with permission of the Heating and Ventilating Contractors' Association.

As indicated in Figure 8.2 the area of maximum wind pressure occurs at around 2/3 of the height of the building. Figure 8.6 shows a typical pressure distribution on a vertical wall facing the wind. Invariably, a building is not located in isolation but forms part of a group of buildings which can vary in density and relative position. Figure 8.7 shows wind flow patterns around a group of buildings and Figure 8.8 shows the wind flow patterns over the

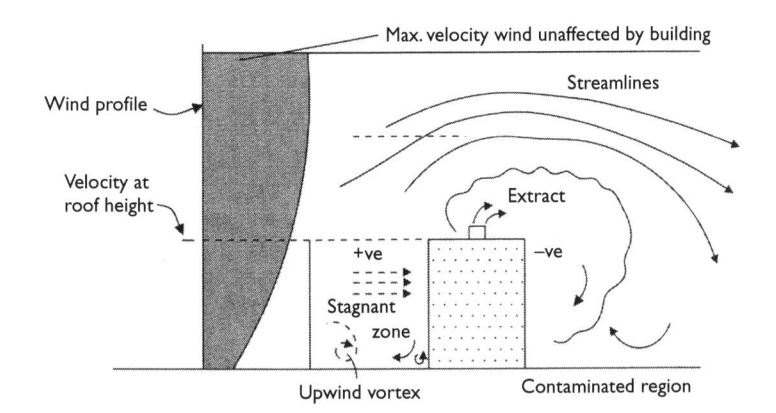

Figure 8.5 Flow patterns over a building showing effect of building height and pollution emission.

Source: Reproduced with permission of the Heating and Ventilating Contractors' Association.

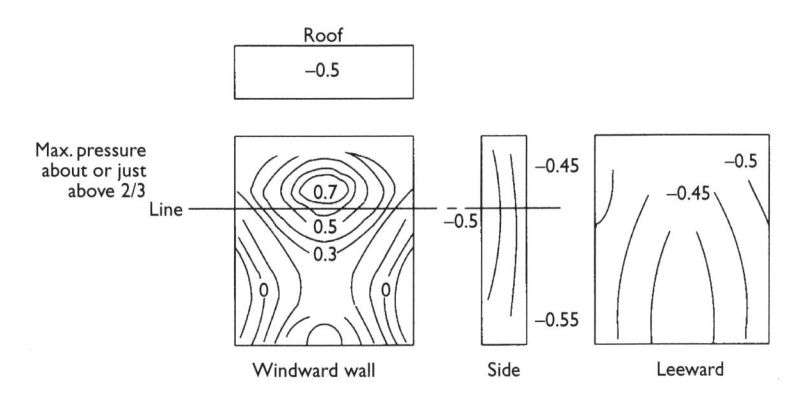

Figure 8.6 Typical pressure distribution for an average building form. (Greatest pressure is at two-thirds point on windward wall.)

Source: Reproduced with permission of the Heating and Ventilating Contractors' Association.

same group of buildings. The zones of positive and negative pressures are identified as well as the potential zone where pollution may be a cause for concern. The wind flow patterns and the effects that groups of buildings have in the path of the wind can be developed using models in a wind tunnel. The factors which influence the effects of wind are

- building shape, size and orientation;
- location of the building with respect to other properties, including their shapes;

190 Natural ventilation in buildings

- natural and artificial wind brakes;
- type of terrain;
- wind speed and direction;
- height above sea level.

It is apparent from the foregoing that the effects of wind on the natural ventilation occurrence in buildings is a complex subject and one which requires the use of the wind tunnel and computer modelling techniques. However, the magnitude and characteristics of natural ventilation can be demonstrated by considering simplified models and by adopting empirical equations.

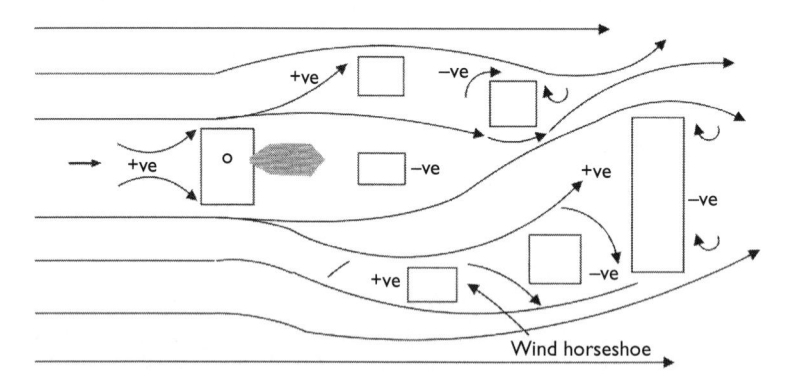

Figure 8.7 Plan of building complex.
Source: Reproduced with permission of the Heating and Ventilating Contractors' Association.

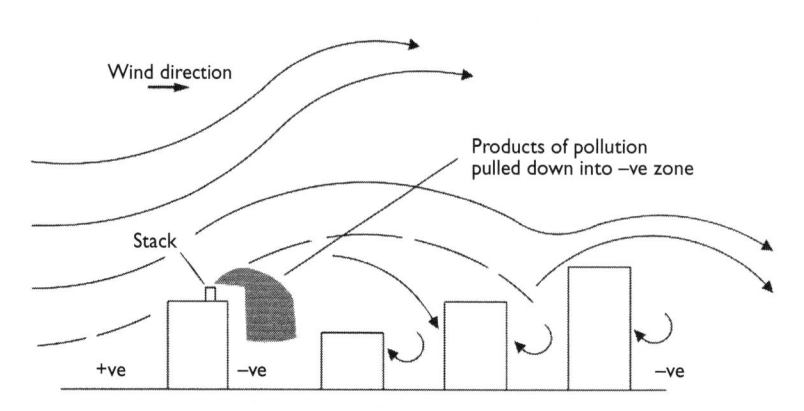

Figure 8.8 Elevation of building complex.
Source: Reproduced with permission of the Heating and Ventilating Contractors' Association.

8.3 Effects on cross-ventilation from the wind

Wind pressure which results from its velocity can be obtained from the velocity pressure term of the Bernoulli equation referred to in Section 7.2, Chapter 7. Thus,

$$P_u = 0.5\rho u^2 \text{ Pa}$$

If the initial wind velocity is u and the final wind velocity is zero

$$dP_u = 0.5\rho u^2 \text{ Pa}$$

Example 8.1 Calculate the wind pressure on a building facade

Determine the pressure caused by the following wind speeds on the facade of a building: 20 km/h, 40 km/h, 80 km/h. Take air density as 1.2 kg/m^3.

Solution
Air speed in m/s = 20 × 1000/3600 = 5.556 m/s, 11.11 m/s and 22.22 m/s.

$$dP_u = 0.5 \times 1.2 \times (5.556)^2 = 18.52 \text{ Pa}, 74.1 \text{ Pa and } 296 \text{ Pa}$$

You will notice that since $dP_u \propto u^2$, as wind speed doubles so velocity pressure quadruples. It is also important to appreciate that, for example, 296 Pa = 296 N/m^2 of facade and although the pressure on the facade of a building is not consistent (see Figure 8.6), the gross lateral force F in Newtons on a building facade measuring 10 m by 15 m high and subject to a wind speed of 80 km/h (50 mph) will be:

F = pressure in Pascals times surface area A in m^2,

$$\text{Pa} \times A = 296(10 \times 15) = 44\,400 \text{ N}$$

This is equivalent to a gross lateral load of 44 400/9.81 = 4526 kg or 4.526 tonne.

Air flow through openings

The rate of air flow subject to wind through an opening is expressed as:

$$Q_w = A_w \cdot C_d (2 \cdot dP/\rho)^{0.5} \text{ m}^3/\text{s}$$
$$\text{or } Q_w = A_w \cdot C_d \cdot u_m (dC_p)^{0.5} \text{ m}^3/\text{s}$$

If the openings are in series as shown in Figure 8.9,

$$1/A_w^2 = [1/(A_1 + A_2)^2] + [1/(A_3 + A_4)^2]$$

Natural ventilation in buildings

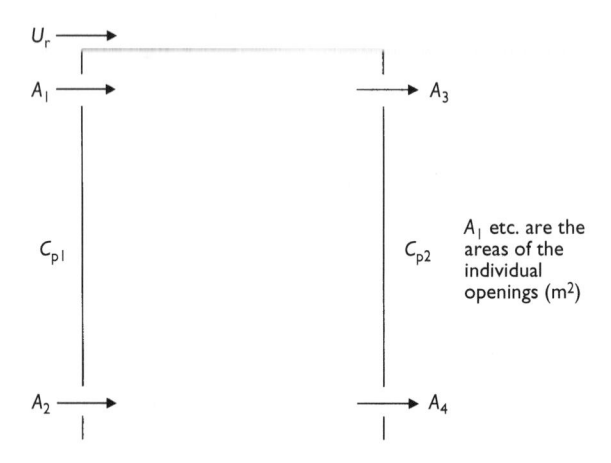

Figure 8.9 Cross-ventilation of a simple building due to wind forces only.

Source: Reproduced with permission of the Heating and Ventilating Contractors' Association.

A_1 etc. are the areas of the individual openings (m²)

Example 8.2 Determine the minimum ventilation rate due to wind speed

(a) Determine the minimum ventilation rate in a workshop due to a mean wind speed of 9 m/s on one of its facades in which there are two openings. The lower opening has a free area of 0.8 m² and the upper opening a free area of 0.3 m². Assume that there are similar openings on the opposite facade. Take air density as 1.2 kg/m³, $dC_p = 1.0$ and $C_d = 0.61$.

(b) If the workshop measures 50 m × 28 m × 5 m high determine the infiltration rate of air resulting from the wind.

Solution (a)
For openings in series

$$1/A_w^2 = [1/(0.8 + 0.3)^2 + 1/(0.8 + 0.3)^2]$$

From which

$$1/A_w^2 = 0.8264 + 0.8264 = 1.653$$

and

$$A_w^2 = 0.605$$

therefore

$$A_w = 0.778 \, m^2$$

and

$$Q_w = A_w \cdot C_d \cdot u_m (dC_p)^{0.5}$$

If $dC_p = 1.0$ and $C_d = 0.61$,

$$Q_w = 0.778 \times 0.61 \times 9 \times 1.0$$

and

$$Q_w = 4.27 \text{ m}^3/\text{s}$$

Now adopting the equation

$$Q_w = A_w \cdot C_d (2dP/\rho)^{0.5}$$

If the pressure drop dP across the building is taken as the drop in velocity pressure dP_u where final velocity is taken as zero then dP = dP_u. Thus,

$$dP = dP_u = 0.5\rho u^2 = 0.5 \times 1.2 \times 9^2 = 48.6 \text{ Pa}$$

Substituting:

$$Q_w = 0.778 \times 0.61 \times (2 \times 48.6 / 1.2)^{0.5}$$

Therefore

$$Q_w = 4.27 \text{ m}^3/\text{s}$$

Solution (b)
Air change rate N can be obtained from

$$Q_w = N \cdot V/3600 \text{ m}^3/\text{s}$$

Rearranging:

$$N = 3600 \cdot Q_w/V$$
$$= 3600 \times 4.27/(50 \times 28 \times 5)$$
$$= 2.2 \text{ air changes per hour}$$

SUMMARY FOR EXAMPLE 8.2
The wind speed is quite high; 9 m/s is equivalent to 20 mph. This is the reason for the high rate of air change. Note that no account has been taken of natural ventilation due to temperature difference between indoors and outdoors.

194 Natural ventilation in buildings

If the wind speed is reduced to 3 m/s which is a more normal value for a less exposed site. The volume flow rate Q_w attributable to wind speed will be 1.424 m³/s and the air change rate N will be 0.732 per hour. You should now confirm these solutions.

8.4 The stack effect

The difference in temperature between inside a building and outside creates thermal forces called stack effect. The more extreme the temperature difference, the greater is the potential for outdoor air to enter the building thus forcing the warm air inside outside. The resulting stack effect is caused by the difference in density between indoor air and air outdoors and the effect is most noticeable during the winter when the greatest temperature difference will be apparent for a heated building.

Figure 8.10 illustrates the stack effect by showing the column of warm air inside the building and a corresponding column of cold air outside. The pressure at the base of the column of outdoor air will be

$$P_o = h \cdot \rho_o \cdot g \text{ Pa}$$

and the pressure at the base of the column of warm air indoors

$$P_i = h \cdot \rho i \cdot g \text{ Pa}$$

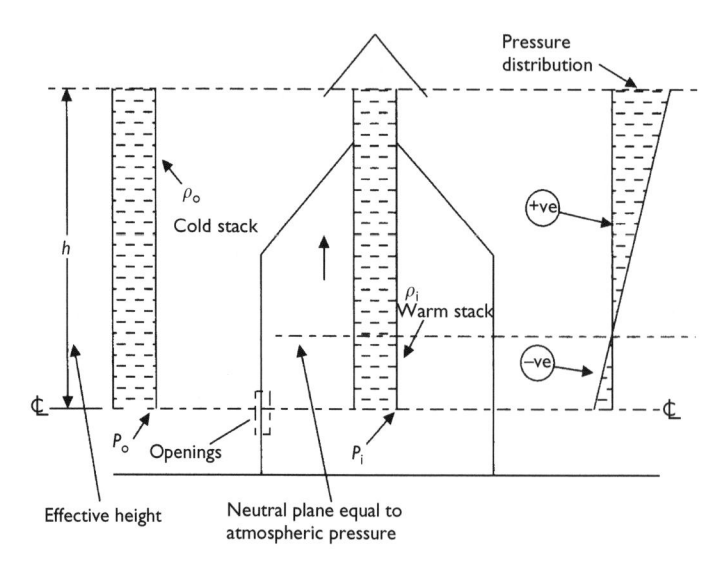

Figure 8.10 Thermal forces (stack effect).

Source: Reproduced with permission of the Heating and Ventilating Contractors' Association.

The air densities are taken as mean values over the height of the column h. The pressure difference dP provides the driving force for air movement from indoors to outdoors. Thus,

$$\mathrm{d}P = h \cdot \mathrm{d}\rho \cdot g \text{ Pa}$$

Figure 8.10 also shows the pressure distribution such that at low level the air movement into the building is subject to suction or negative pressure (−ve) due to its buoyancy and the exit point of the air from the building at high level is subject to positive pressure (+ve). The neutral point occurs at a horizontal plane where the negative air pressure changes to positive pressure. The neutral point is at atmospheric pressure.

There are two methodologies for the determination of the stack effect.

1. Air density at 20°C and 101 325 Pa is 1.2 kg/m^3 and this frequently is taken as the mean density of the air indoors. Since air density is inversely proportional to its absolute temperature and atmospheric pressure is considered constant both indoors and outdoors, the density of outdoor air can then be obtained from $\rho_o = \rho_i(T_i/T_o)$. From which $\mathrm{d}P = h \cdot \mathrm{d}\rho \cdot g$ Pa.

   Alternatively, air densities can be obtained from the *Thermodynamic and Transport Properties of Fluids*.

2. However, it is more convenient to determine the pressure drop caused by the stack effect from knowledge of the mean temperature of each column of air. From the equation $\mathrm{d}P = h \cdot \mathrm{d}\rho \cdot g$, the pressure difference can be expressed as:

$$\mathrm{d}P = h \cdot g(\rho_o - \rho_i) \text{ Pa}$$

The density of outdoor air at 0°C, 273 K is 1.293 kg/m^3. This air density can be put in the form

$$\rho_o = (1.293 \times 273)/(273 + t_o)$$

The density of the air indoors will therefore be

$$\rho_i = (1.293 \times 273)/(273 + t_i)$$

Substituting these two equations into the formula for dP:

$$\mathrm{d}P = h \cdot g[(1.293 \times 273)/(273 + t_o) - (1.293 \times 273)/(273 + t_i)]$$

thus

$$\mathrm{d}P = (1.293 \times 273)h \cdot g[(1/(273 + t_o)) - (1/(273 + t_i))]$$

from which

$$dP = 3463h[(1/(273 + t_o)) - (1/(273 + t_i))]$$

There now follows an example in which this equation for stack effect is employed.

Example 8.3 Calculate the stack effect in a multi-storey building

A building is 15 storeys high and is held at a temperature of 20°C. Determine the potential stack effect when outdoor temperature is −4°C, given that the floor to ceiling height is 3 m.

Solution
Assuming that the stack effect extends to the full height of the building

$$dP = 3463 \times 15 \times 3[(1/268) - (1/293)] = 49.6 \text{ Pa}$$

Now try using $dP = h \cdot d\rho \cdot g$ Pa.

Cross-ventilation through openings

The rate of air flow subject to stack effect through an opening is:

$$Q_s = C_d \cdot A_s[(2dt \cdot h \cdot g)/(t_m + 273)]^{0.5} \text{ m}^3/\text{s}$$

For apertures in series (as shown in Figure 8.11),

$$1/A_s^2 = [(1/(A_1 + A_3))^2 + (1/(A_2 + A_4))^2]$$

Note the difference in this equation for A_s compared with the equation for A_w.

Example 8.4 Determine the ventilation rate due to stack effect

Determine the ventilation rate due to stack effect for the building in Example 8.3 given that the lower and upper openings on one facade are 0.5 and 0.3 m² with similar openings on the facade opposite to it. Take C_d as 0.61.

Solution
Substituting we have

$$1/A_s = [(1/(0.8)^2) + (1/(0.8)^2)]$$

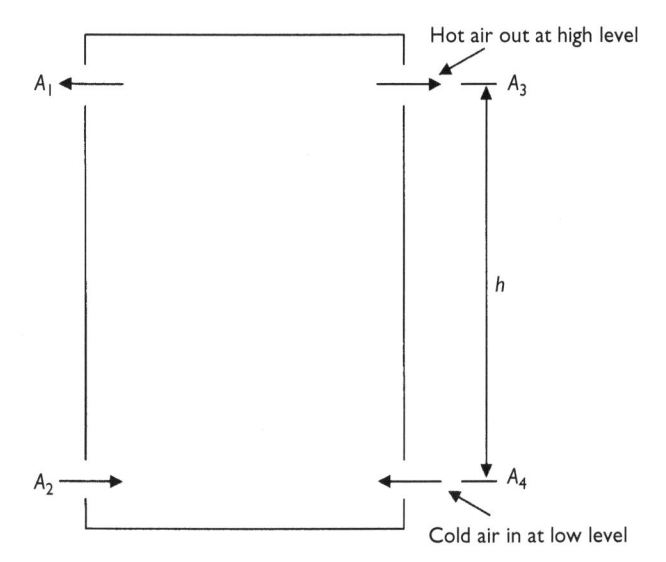

Figure 8.11 Cross-ventilation of a simple building due to temperature difference only.
Source: Reproduced with permission of the Heating and Ventilating Contractors' Association.

from which $A_s = 0.32$ m². Substituting we have

$$Q_s = 0.61 \times 0.32[(2(20+4)(3 \times 15) \times 9.81)/(8+273)]^{0.5}$$
$$= 1.695 \text{ m}^3/\text{s}$$

Note: t_m is the mean of 20°C and −4°C which is 8°C since the difference between 20 and 8 is 12 K and the difference between 8 and −4 is 12 K.

SUMMARY FOR EXAMPLES 8.2 AND 8.4
These examples deal with uninhibited cross-ventilation where there are no internal partitions. For these simple applications the actual ventilation rate may be taken as the larger of that due to the wind or stack effect. It is likely that in the summer the building will be subject to moderate natural ventilation since wind speed will normally be low with a small difference between indoor and outdoor temperature.

If the building is air conditioned with indoor temperature lower than outdoor temperature during the summer, the stack effect will be reversed. This means that the cooler air from the building will emanate form openings at low level.

8.5 Natural ventilation to internal spaces with openings in one wall only

Figure 8.12 refers to the effect of wind incident upon a facade having one opening where the approximate volume flow

$$Q_w = 0.025A \cdot u_h \text{ m}^3/\text{s}$$

The velocity of the wind tends to increase with height above ground level. See Figure 8.5. Figure 8.13 refers to the effect of indoor to outdoor

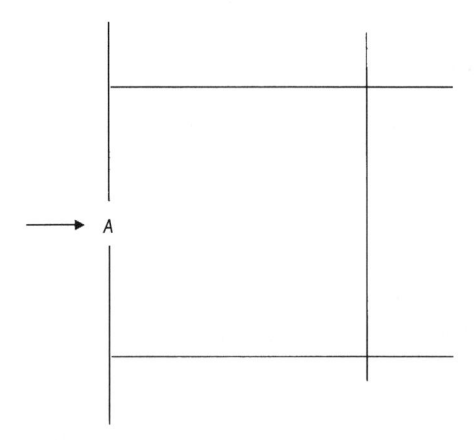

Figure 8.12 Internal space subject of wind.

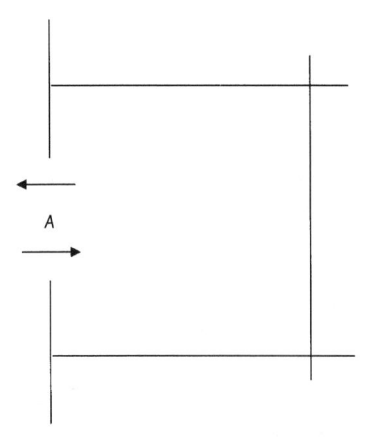

Figure 8.13 Internal space subject to temperature difference.

Natural ventilation in buildings 199

Table 8.1 Values of J for two types of window

Angle of opening	Type of window	J
30	Side mounted casement	0.6
60	Side mounted casement	0.9
90	Side mounted casement	1.1
30	Centre pivoted window	0.7
60	Centre pivoted window	0.92
90	Centre pivoted window	0.95

Source: Reproduced from *CIBSE Guide* section A4 (1986) by permission of the Chartered Institution of Building Services Engineers.

temperature difference for one opening in the facade where the approximate volume flow

$$Q_s = C_d(A/3)J[(dt \cdot h \cdot g)/(t_m + 273)]^{0.5} \text{ m}^3/\text{s}$$

The value of J depends upon the angle of opening for the window. Table 8.1 lists some typical values.

Example 8.5 Calculate the volume flow of air into an internal space due to an opening on the windward side

Determine the volume flow rate of outdoor air into an internal space having an opening on the windward side whose equivalent area is 6400 mm². The mean wind speed at the building height of 10 m is 9 m/s.

Solution
Substituting into the formula we have,

$$Q_w = 0.025 \times (6400 / 1\,000\,000) \times 9 = 0.00144 \text{ m}^3/\text{s}$$
$$= 1.44 \text{ L/s}$$

Example 8.6 Determine the volume flow and air change into a room from a partially open casement window

(a) Determine the volume flow rate of air exchange through a side mounted casement window to an internal space, given that the angle of opening is 60°. Indoor temperature is 25°C and outdoor temperature is 10°C and the height of the casement is 0.8 m by 0.5 m wide.
(b) Determine the air change rate for the room as a consequence of the opened window, given that it measures 5 × 4 × 2.7 m high.

200 Natural ventilation in buildings

Solution (a)
The value of J, from Table 8.1 is 0.9 and $t_m = (25+10)/2 = 12.5°C$. Substituting the data into the equation for Q_s, for natural ventilation due to temperature difference:

$$Q_s = (0.61 \times (0.8 \times 0.5)/3) \times 0.9[(25-10) \times 0.8 \times 9.81/(12.5+273)]^{0.5}$$

$$= 0.047 \text{ m}^3/\text{s}$$

$$= 47 \text{ L/s}$$

Solution (b)
From the equation, $Q = N \cdot V/3600 \text{ m}^3/\text{s}$, the air change rate $N = 3600 \cdot Q/V = 3600 \times 0.047/(5 \times 4 \times 2.7) = 3.13$ per hour.

8.6 Ventilation for cooling purposes

Current design favours the use of natural or fan assisted ventilation of the building shell for maintaining comfort conditions in preference to the use of air conditioning plant. The building envelope must be designed and orientated in order to take advantage of the wind and the stack effect caused by indoor to outdoor differences in temperature. A number of articles relating to buildings designed in this manner have appeared in the monthly journal of the *CIBSE*. Refer to Section 8.8.

Where an air conditioning plant is required, night-time cooling of the building shell using natural ventilation can also be made advantageous. Again, the design of the building must account for the flow paths for the ventilating outdoor air to ensure cooling of the building structure and exfiltration of the resultant warmed air. This involves analysis of the building's thermal response to summer outdoor temperatures and solar heat gains to ensure a thermally stabilised building structure at the commencement of occupation at the beginning of the day.

Figure 8.14 shows a section through a building designed for cooling by ventilation.

Case study 8.1 Size the free area for openings, estimate the rate of cooling and daily cooling energy extraction for a five-storey building

A five-storey building measures 30×15 and has floor to ceiling heights of 3 m. It is to be structurally cooled in the summer months during the evening and night by natural ventilation to provide a total of 42 air changes during the unoccupied period which extends from 1800 hours to 0800 hours.

Natural ventilation in buildings 201

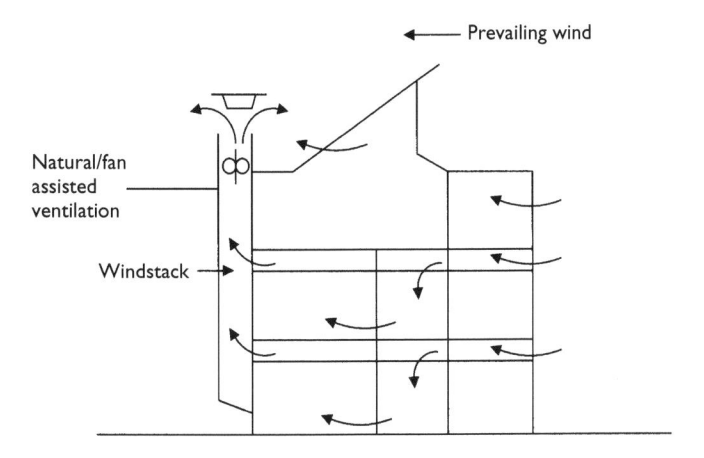

Figure 8.14 Section through a building designed for cooling by ventilation.

Determine the free area of each of the two openings on the leeward side of the building.

(a) Given design mean indoor and outdoor air temperatures during the unoccupied period are 25°C and 15°C respectively.
(b) For a wind speed of 5m/s.

Data: The two openings on the windward side of the building have free area of 1.5 m² at low level and 1.8² at high level. Take $C_d = 0.61$, $h = 13.5$ m. Assume the openings are in series.

(c) Estimate the rate of free cooling in kW from the natural ventilation of the building when indoor temperature is 25°C and outdoor temperature is 15°C. Take air density as 1.2 kg/m³.
(d) Estimate the daily cooling energy in kWh extraction by natural ventilation during the hours of 1800 and 0800.

Solution (a)
The required air change rate per hour = 42/(6p.m. to 8a.m.) = 42/14 = 3
From $Q = NV/3600$
$\qquad = 3 \times (30 \times 15 \times 3 \times 5)/3600 = 5.625$ m³/s

Estimating for temperature difference

$$Q_s = C_d \cdot A_s[(2dt \cdot h \cdot g)/(t_m + 273)]^{0.5} \text{ m}^3/\text{s}$$

where $t_m = (15 + 25)/2 = 20°C$. Substituting

$$5.625 = 0.61 A_s[(2(25 - 15) \times 13.5 \times 9.81)/(20 + 273)]^{0.5}$$

from which

$$5.625 = 0.61 A_s \times 3.0066$$

and

$$A_s = 3.07 \text{ m}^2$$

For apertures in series

$$1/A_s^2 = [(1/(A_1 + A_3)^2) + (1/(A_2 + A_4)^2)]$$

Substituting

$$1/3.07^2 = [(1/(1.8 + A_3)^2) + (1/(1.5 + A_4)^2)]$$

Assuming

$$A_3 = A_4 = A$$

$$0.1061 = [(1/(1.8 + A)^2) + (1/(1.5 + A)^2)]$$

To solve the equation it can be expressed as $z = x + y$ where $z = 0.1061$ and with values allocated for A and the free area of openings A_3 and A_4.

The solution is tabulated and given in Table 8.2. From Table 8.2 the free area for apertures A_3 and A_4 is approximately 2.7 m² each since z is almost equal to 0.1061.

Solution (b)
For ventilation resulting from wind $Q_w = C_d \cdot A_w (2dP/\rho)^{0.5}$ m³/s. Now $dP = dP_u = 0.5\rho u^2$. At a mean temperature $t_m = 20°C, \rho = 1.2$ kg/m³. Thus,

$$dP = 0.5 \times 1.2 \times 5^2 = 15 \text{ Pa}$$

and

$$Q_w = 0.61 A_w (2 \times 15/1.2)^{0.5} = 3.05 A_w$$

Table 8.2 Solution to openings for free areas A_3 and A_4 Case study 8.1a

A	x	y	z
2	0.06925	0.8163	0.1509
2.5	0.05408	0.0625	0.1166
2.7	0.04938	0.05669	0.10607

Substituting for Q_w,

$$5.625 = 3.05 A_w$$

from which

$$A_w = 1.844 \text{ m}^2$$

For apertures in series

$$1/A_w^2 = [(1/(A_1 + A_2)^2) + (1/(A_3 + A_4)^2)]$$

substituting

$$1/(1.844)^2 = [(1/(1.5 + 1.8)^2) + (1/(A_3 + A_4)^2)]$$

If $A_3 = A_4$,

$$1/3.4 = [(1/10.89) + (1/(2A)^2)]$$

rearranging

$$0.294 - 0.0918 = 1/(2A)^2$$

from which

$$0.2022 = 1/(2A)^2$$

and

$$4.946 = (2A)^2$$

thus

$$2.224 = 2A$$

and

$$A = 1.112 \text{ m}^2$$

Therefore, the free area of each of the apertures A_3 and A_4 is 1.112 m^2.

Solution (c)
The building structure will absorb heat by solar radiation and conduction during the hours of daylight. With the ventilation system in use during occupation, this will help to ensure against excessive indoor temperatures.

The mass transfer of air through the building by natural ventilation

$$M = Q \cdot \rho = 5.625 \times 1.2 = 6.75 \text{ kg/s}$$

The maximum rate of cooling between 1800 hours and 0800 hours can be determined from: rate of cooling $= M \cdot C \cdot dt$ kW; where for air, $C = 1.025$ kJ/kgK. Substituting:

the rate of free cooling $= 6.75 \times 1.025 \times (25 - 15) = 69$ kW

Solution (d)
The estimated daily cooling energy extracted by natural ventilation can be determined by taking log mean temperature difference (LMTD) between indoors and outdoors between 1800 and 0800 hours. LMTD accounts for there being a change both in outdoor temperature and a change in indoor temperature and the need to find the true mean difference between these two temperature changes. See Section 3.3.

At the beginning of the unoccupied period when cooling of the building is considered at 1800 hours, it is possible that indoor temperature and outdoor temperature will be equal at 25°C and no cooling takes place. At some point during the night, outdoor temperature drops to 15°C and with indoor temperature still at or near 25°C, the maximum cooling rate will be 69 kW and the maximum temperature difference is $(25 - 15) = 10$ K. The minimum temperature difference will occur at 0800 hours when indoor temperature will be about 18°C and outdoor temperature 15°C. Thus minimum temperature difference is $(18 - 15) = 3$ K

The LMTD will be

$$\text{LMTD} = (dt_{max} - dt_{min})/\ln(dt_{max}/dt_{min})$$

Thus LMTD $= (10 - 3)/\ln(10/3) = 5.814$ K.

Note: The *arithmetic* (less accurate) mean temperature difference $= (10 + 3)/2 = 6.5$ K. Now the daily energy extracted

$$H = M \cdot C \cdot dt \times \text{time kWh}$$

The units of the terms are:

$$H = (\text{kg/s})(\text{kJ/kgK})(\text{K})(\text{h}) = (\text{kJ/s})(\text{h}) = \text{kWh}$$

Substituting:

daily energy extracted $= 6.75 \times 1.025 \times 5.814 \times (1800 - 0800)$
$= 563$ kWh

SUMMARISING THE SOLUTIONS TO CASE STUDY 8.1
Maximum rate of cooling by natural ventilation at night = 69 kW; estimated daily cooling energy from natural ventilation = 563 kWh.

Conditions	A_1	A_2	A_3	A_4
Stack effect, $t_i = 25°C, t_o = 15°C$	1.8 m²	1.5 m²	2.7 m²	2.7 m²
Wind effect, $u = 5$ m/s	1.8	1.5	1.112	1.112

There is a substantial difference in free area for apertures A_3 and A_4 between the stack effect and the effect of the wind. You should now consider the effect of a wind speed of 3 m/s upon size of apertures A_3 and A_4 as air movement at night can be quite low in the summer.

The prevailing conditions will be a combination of wind and stack effect. However, the larger apertures determined from either wind or temperature difference would be considered appropriate.

QUALIFYING REMARKS RELATING TO CASE STUDY 8.1
There have been a number of assumptions made in the solution to case study 8.1 and the following qualifying remarks must be made.

- It is assumed that cross-ventilation takes place with no internal partitions.
- Recourse should be made to establish minimum summertime outdoor temperatures which normally occur at night time. This will depend upon geographical location.
- The building's thermal capacity and orientation, including the ventilation pathways, need analysing to ensure that peak indoor temperature normally occurs at the end of, or after, the occupation period of 1800 hours.
- Peak indoor temperature will also need to be set as a design parameter in the modelling process.
- Four apertures in series have been considered, two on the windward and two on the leeward side of the offices.

8.7 Fan assisted ventilation

If the building is located in a sheltered position, the use of extract fans can provide a positive air displacement for the building. They can also be used to advantage when the indoor to outdoor temperature difference is small thus reducing the influence of the stack effect. The extract fans can be controlled by wind speed and direction and indoor temperature so that they are only used when necessary to aid in capturing the heat energy absorbed by the building during the day. Refer again to Figure 8.14.

8.8 Further reading

Building Services, the *CIBSE Journals* (*BSJ*):

November, 1994. De Montfort University School of Built Environment.
December, 1994. Advancing the Cause of Low Energy Design.
February, 1995. Which Ventilation system?
April, 1995. East Anglia University New Academic Building.
May, 1995. BRE Low Energy Office.
June, 1995. Control Naturally. BSRIA Research Project.
November, 1995. No.1 Leeds City Office Park.
January, 1996. Canterbury Court Centre.
February, 1996. Ventilation Solutions
March, 1996. Future Buildings.
April, 1996. Green Buildings; Benefits and Barriers.
June, 1996. Cable and Wireless College, Probe 5.
August, 1996. Probe 6, Woodhouse Medical Centre.
November, 1996. Refurbishment the Natural Way.
December, 1996. Probe 8, Queen's Building, Anglia Polytechnic University. Probe 8.
January, 1997. Salt-bath Modelling of Air Flows.
February, 1997. Portsmouth University, Portland Building.
March, 1997. Green Demo, Environmental Office of the Future.

The *BSJ* in subsequent monthly publications regularly includes examples of naturally ventilated buildings that avoid the use of air conditioning.

8.9 Chapter closure

You now have knowledge of the forces and factors affecting the natural ventilation of buildings with respect to stack effect and wind. You have the skills required to undertake simple modelling processes relating to the size and location of apertures in the building envelope and to the mass transfer of air through the building. An approximate methodology to estimate the cooling effect of night time ventilation of the building structure has been investigated. For realistic assessment of natural ventilation as a means of structurally cooling a building at night, recourse must be made to computerised modelling techniques that are beyond the scope of this publication.

Chapter 9

Regimes of fluid flow in heat exchangers

Nomenclature

A	heat exchange surface (m^2)
C_c	specific heat capacity of cold fluid (kJ/kgK)
C_h	specific heat capacity of hot fluid (kJ/kgK)
CR	capacity ratio
d	diameter (m)
dt	temperature difference (K)
dt_m	true mean temperature difference (K)
dt_{max}	maximum temperature difference (K)
dt_{min}	minimum temperature difference (K)
E	effectiveness
exp.	exponential
f	correction for cross flow
h	specific enthalpy of the superheated vapour (kJ/kg)
h_f	specific enthalpy of the saturated liquid (kJ/kg)
h_{fg}	latent heat of evaporation (kJ/kg)
h_{si}	inside heat transfer coefficient (kW/m^{2}K)
h_{so}	outside heat transfer coefficient (kW/m^{2}K)
HVAC	heating ventilating and air conditioning
h_w	specific enthalpy of the wet vapour (kJ/kg)
HWS	hot water service
k	thermal conductivity (kW/mK)
L	length (m)
LMTD	log mean temperature difference (K)
LTHW	low temperature hot water space heating
M_c	mass transfer of cold fluid (kg/s)
M_h	mass transfer of hot fluid (kg/s)
NTU	number of transfer units
q	dryness fraction

Q	output (kW)
r_1, r_2	radius for radial heat transfer (m)
Re	Reynolds number
R_f	fouling thermal resistance (m²K/W)
R_t	total thermal resistance (m²K/kW)
t_{c1}, t_{c2}	inlet and outlet temperatures of the secondary fluid (°C)
t_{h1}, t_{h2}	inlet and outlet temperatures of the primary fluid (°C)
t_m	mean temperature (°C)
U	overall heat transfer coefficient (kW/m²K)
U_L	overall heat transfer coefficient (kW/mK)
Z, Z_1	temperature ratios for cross flow

9.1 Introduction

There are many different types of heat exchangers available to the building services industry. Plate heat exchangers and heat pipes which have been in use in other industries for many years are now used in heat ventilating and air conditioning (HVAC) systems for extracting low grade heat from return air in ventilation and air conditioning systems, for example. The thermal wheel is also used for this purpose. The cooling tower employed for cooling condenser water in an air conditioning plant allows the two fluids, atmospheric air and condenser water, to come into direct contact for heat transfer to take place.

However, this chapter focuses on heat exchangers having a solid boundary between the two fluids. The function of this type of heat exchanger is to allow the transfer of heat energy between two fluids at different temperatures across the solid boundary. It is used to ensure that the two fluids do not come into direct contact. An ideal heat exchanger of this type should achieve maximum rate of heat exchange using the minimum heat exchange space and the minimum pressure drop on both sides of the solid boundary.

In practice, if the solid boundary is a plain straight tube, a comparatively large heat exchange space will required although the pressure drop will be relatively low. Alternatively, a solid boundary in the form of a coiled tube with finning extends the heat transfer surface in a small space but the pressure loss inside and outside the tube bundle will be comparatively high. Inevitably, a compromise is usually made in heat exchanger design for specific applications.

Table 9.1 lists some of the heat exchangers used in the Building Services industry. You should familiarise yourself with the construction of the various heat exchangers on the market from manufacturers' current literature.

9.2 Parallel flow and counterflow heat exchangers

Figure 9.1 shows a parallel flow heat exchanger with its accompanying temperature distribution assuming both fluids vary in temperature. t_{h1} being the initial temperature of the hot fluid and t_{c1} the initial temperature of the cold fluid.

Table 9.1 Examples of heat exchangers in the Building Services industry

Heat exchanger	Media
Double pipe	Water
Shell and tube	Water/condensing/evaporating fluids
Plate	Water/air, water/water
Run around coils	Water/air
Pipe coils	Water/air
Heat pipes	Water/air
Regenerator (thermal wheel)	Air/air
Spray condenser/de-superheater/flash steam recovery/cooling tower	Water/air, condense/steam, air/water

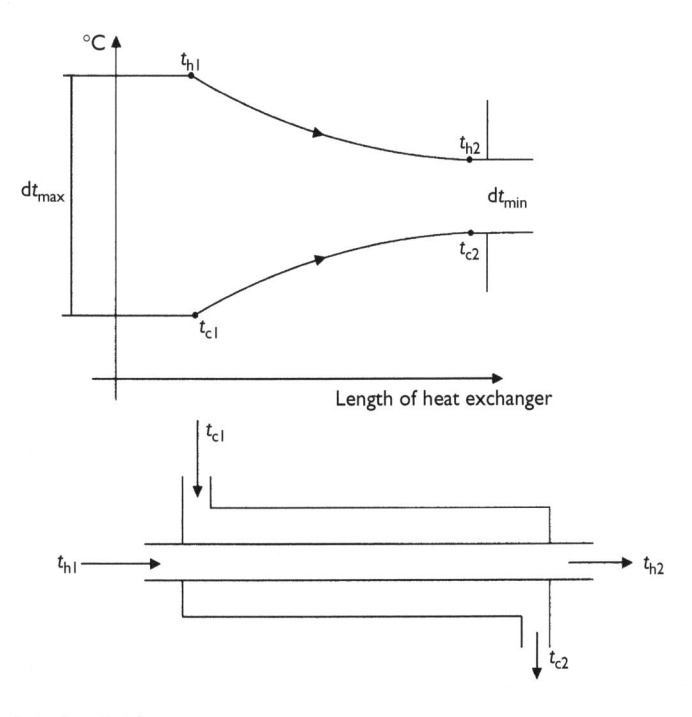

Figure 9.1 Parallel flow.

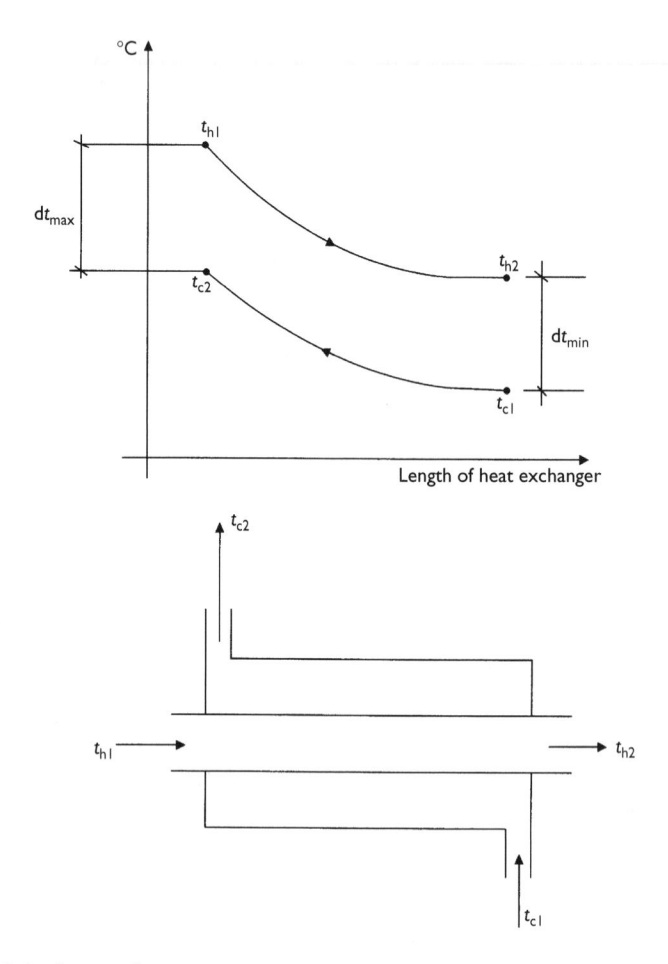

Figure 9.2 Counterflow.

Figure 9.2 shows a heat exchanger in counterflow with its accompanying temperature distribution assuming both fluids vary in temperature on their passage through the heat exchanger. Again t_{h1} is the initial temperature of the hot fluid and t_{c1} the initial temperature of the cold fluid.

For practical reasons, for a heat exchanger in parallel flow the final temperatures t_{h2} and t_{c2} can never be equal and clearly t_{c2} cannot exceed t_{h2}. However, for a heat exchanger in counterflow t_{c2} can exceed t_{h2}. Refer to Figure 9.2. Therefore, parallel flow has a limitation on the relationship between the primary and secondary leaving temperatures.

In cases where both the primary and secondary fluids vary in temperature the arithmetic mean temperature difference which provides the motive force

Regimes of fluid flow in heat exchangers 211

in the heat exchange does not always register the true mean temperature difference and the log mean temperature difference (LMTD) between the two fluids is adopted.

$$\text{LMTD } dt_m = (dt_{max} - dt_{min}) / \ln(dt_{max}/dt_{min}) \text{ K}$$

If, however, $dt_{max} = dt_{min}$, LMTD dt_m = zero and the arithmetic mean temperature difference is used.

For heat exchangers in parallel flow,

primary temperatures $t_{h1} \rightarrow t_{h2}$

secondary temperatures $t_{c1} \rightarrow t_{c2}$

From which $dt_{max} = t_{h1} - t_{c1}$ and $dt_{min} = t_{h2} - t_{c2}$. For heat exchangers in counterflow

primary temperatures $t_{h1} \rightarrow t_{h2}$

secondary temperatures $t_{c2} \leftarrow t_{c1}$

From which $dt_{max} = t_{h1} - t_{c2}$ and $dt_{min} = t_{h2} - t_{c1}$. For both types of flow dt_{max} and dt_{min} can be reversed.

Example 9.1 Find the log mean temperature difference

Determine the true temperature difference for the primary and secondary fluids for a heat exchanger in counterflow.

(a) Primary fluid inlet temperature = 120°C, outlet temperature = 90°C. Secondary fluid inlet temperature = 10°C, outlet temperature = 80°C.
(b) Primary fluid inlet temperature = 100°C, outlet temperature = 80°C. Secondary fluid inlet temperature = 10°C, outlet temperature = 80°C.

Solution (a)

(a) Primary temperatures   120 → 90
    Secondary temperatures 80 ← 10
    $dt_{max,min}$ $dt_{min} 40 - dt_{max} 80$
    and LMTD $dt_m = (80 - 40)/\ln(80/40) = 57.7$ K

(b) Primary temperatures   100 → 80
    Secondary temperatures 30 ← 10
    $dt_{max,min}$ $dt_{min} 70 - dt_{max} 70$
    and LMTD dt_m = zero

212 Regimes of fluid flow in heat exchangers

The arithmetic mean temperature difference must therefore be used here and will be:

Primary mean temperature $(100 + 80)/2 = 90$

Secondary mean temperature $(30 + 10)/2 = 20$

Arithmetic mean temperature difference $dt_m = 90 - 20 = 70$ K

Boiling and condensing in parallel and counterflow

It is possible for one of the fluids to remain at constant temperature during the process of heat exchange across the solid boundary. The evaporator is an example in which the fluid being cooled causes the cool fluid to evaporate at constant temperature. Figure 9.3 shows the heat exchanger and the accompanying process.

The condenser is another example in which the fluid being condensed at constant temperature causes the cool fluid to rise in temperature. Figure 9.4 shows the heat exchanger and the accompanying process.

9.3 Heat transfer equations

The rate of heat transfer may be expressed in various ways:

$$Q = U \cdot A \cdot dt_m \text{ kW}$$

where $U = 1/R_t$ kW/m^{2}K and $R_t = (1/h_{si}) + R_f + (1/h_{so})$ m^{2}K/kW. h_{si} and h_{so} derive from the laminar sublayers on either side of the solid boundary, see Chapter 6. The thermal resistance of the solid boundary is not significant and is, therefore, sometimes ignored.

$$Q = M_c \cdot C_c(t_{c2} - t_{c1}) \text{ kW}$$
$$Q = M_h \cdot C_h(t_{h1} - t_{h2}) \text{ kW}$$
$$Q = M_h \cdot h_{fg} = M_h \cdot (q \cdot h_{fg}) \text{ kW}$$
$$Q = M_h(h - h_f) = M_h(h_w - h_f) \text{ kW}$$

Ignoring the inefficiency of heat exchange, a heat balance may be drawn such that:

heat lost by the primary fluid = heat gained by the secondary fluid

For example:

$$M_h \cdot C_h(t_{h1} - t_{h2}) = M_c \cdot C_c(t_{c2} - t_{c1})$$

from which one unknown can be evaluated.

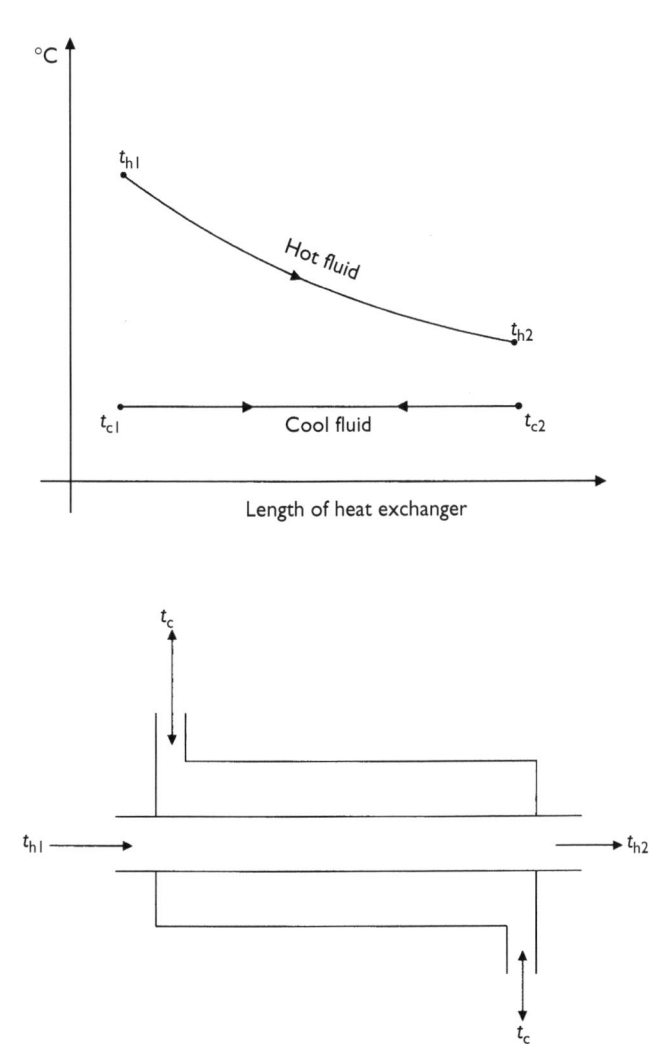

Figure 9.3 Boiling in a shell and tube exchanger, cool fluid boiling.

Fouling factors

The overall heat transfer coefficient U for the solid boundary between the primary and secondary fluids, introduced above, can account for a layer of dirt or scale on the heat exchanger surface in contact with the fluid. Expressed as a thermal resistance R_f to heat flow, it will be at a minimum value at commissioning and reach a maximum resistance at the point when

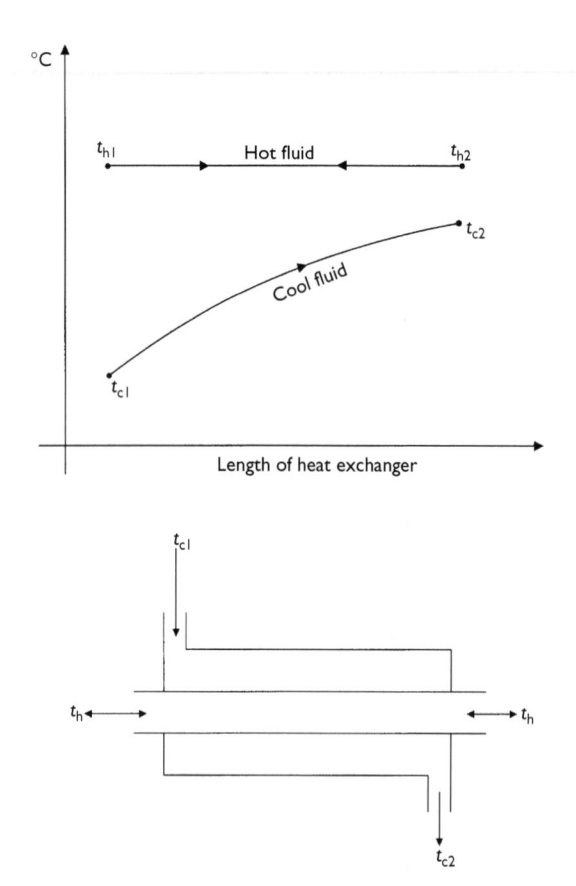

Figure 9.4 Condensing in a shell and tube exchanger, hot fluid condensing.

cleaning and descaling is scheduled. In practice it is difficult to evaluate and depends upon

- fluid properties,
- fluid temperatures,
- fluid velocities,
- heat exchanger material and materials used elsewhere in the system and
- heat exchanger configuration.

Two approaches to accounting for fouling resistance include making a correction to the overall heat transfer coefficient for the heat exchanger or alternatively calculating the fouling resistance from various trials thus:

$$R_f = (1/U_{\text{dirty}}) - (1/U_{\text{clean}})$$

Clearly the effects of scale and dirt, if not dealt with under a regulated planned maintenance, may have a significant effect upon the performance of the heat exchanger. Key issues would include the need for water treatment, inspection, cleaning and flushing out. There now follows two examples relating to the subject matter set out above.

Example 9.2 Compare the advantages of counterflow over parallel flow

A hot fluid is cooled from 118°C to 107°C in a double pipe heat exchanger. Assuming the overall heat transfer coefficient to remain constant, compare the advantage of counterflow over parallel flow in the amount of heat transfer area required when a cold fluid is to be heated from:

(a) 57°C to 104°C.
(b) 30°C to 77°C.
(c) 10°C to 57°C.

Solution
You should note that the temperature rise of the secondary fluid is 47 K in each case. Examples of double pipe heat exchangers are shown in Figures 9.1 and 9.2.

Consider counterflow

	(a)	(b)	(c)
Primary fluid	118 → 107	118 → 107	118 → 107
Secondary fluid	104 ← 57	77 ← 30	57 ← 10
$dt_{min,max}$	14 50	41 77	61 97

LMTD $dt_m = (50 - 14)/\ln(50/14) = 28.3$ K
$dt_m = (77 - 41)/\ln(77/41) = 57.1$ K
$dt_m = (97 - 61)/\ln(97/61) = 77.6$ K

Consider parallel flow

	(a)	(b)	(c)
Primary fluid	118 → 107	118 → 107	118 → 107
Secondary fluid	57 → 104	30 → 77	10 → 57
$dt_{min,max}$	61 3	88 30	108 50

LMTD $dt_m = (61 - 3)/\ln(61/3) = 19.3$ K
$dt_m = (88 - 30)/\ln(88/30) = 53.9$ K
$dt_m = (108 - 50)/\ln(108/50) = 75.3$ K

Since the overall heat transfer coefficient remains constant the true mean temperature differences can now be used to show the advantage of counterflow over parallel flow in relation to the amount of heat transfer surface required.

(a) $28.3/19.3 = 1.466$, (b) $57.1/53.9 = 1.059$, (c) $77.6/75.3 = 1.031$

ANALYSIS OF EXAMPLE 9.2
Although the secondary fluid has the same temperature rise in each case, the advantage of counterflow diminishes as the inlet temperature of the secondary fluid is reduced.

Example 9.3 Determine the heating surface for a shell and tube heat exchanger

A shell and tube heat exchanger is required to raise 0.5 kg/s of low temperature hot water space heating (LTHW) from 70°C to 85°C

(a) Determine the heating surface required if 0.0158 kg/s of steam at 4 bar absolute and 0.9 dry is used as the primary medium in parallel flow.
    Data: Overall heat transfer coefficient $U = 1.2$ kW/m^{2}K; specific heat capacity for water $C = 4.18$ kJ/kgK.
(b) What effect does counterflow have upon the surface area?

Solution (a)
Use will be made of the tables of *Thermodynamic and Transport Properties of Fluids* in the solution.
(a) The output of the heat exchanger can be obtained from the secondary side and

$$Q = M_c \cdot C_c \cdot dt = 0.5 \times 4.18 \times (85 - 70) = 31.35 \text{ kW}$$

Also

$$Q = M_h(h_w - h_f) \text{ kW}$$

and from the tables

$$h_w = (h_f + qh_{fg}) = 605 + 0.9 \times 2134 = 2526 \text{ kJ/kg}$$

Substituting:

$$31.35 = 0.01583(2526 - h_f)$$

From which

$$h_f = 546 \text{ kJ/kg}$$

from the tables therefore,

$$t_s = t_{h2} = 130°C$$

For parallel flow: primary fluid (steam) 143.6 → 130
secondary fluid (water) 70 → 85
$dt_{max,min}$ 73.6   45

LMTD $dt_m = (73.6 - 45)/\ln(73.6/45) = 58$ K

since

$$Q = U \cdot A \cdot dt_m \text{ kW}$$

then heat exchange surface

$$A = Q/U \cdot dt_m = 31.35/1.2 \times 58 = 0.45 \text{ m}^2$$

Solution (b)
Considering counterflow: primary fluid 143.6 → 130
secondary fluid   85 ← 70
$dt_{max,min}$ 58.6   60

Since these mean temperatures are closely similar the arithmetic mean can be taken and:
For the primary fluid,

$$t_m = (143.6 + 130)/2 = 136.8°C$$

For the secondary fluid,

$$t_m = (85 + 70)/2 = 77.5°C$$

from which

$$dt_m = 136.8 - 77.5 = 59.3 \text{ K}$$

You should now determine the LMTD to confirm that it agrees with this arithmetic mean. This closely corresponds to the LMTD of 58 K for parallel flow and therefore will have little influence over the heat exchanger surface.

218 Regimes of fluid flow in heat exchangers

Example 9.4 Calculate the heating surface and output for a vapour compression refrigeration condenser

Tetrafluoroethane (Refrigerant 134a) leaves a compressor at 7.7 bar absolute with 20 K of superheat and enters a condenser at the rate of 0.025 kg/s. The coolant temperature at entry is 12°C at a mass flow rate of 0.08 kg/s. Assuming counterflow, determine the heat exchange surface and the output of the condenser.

Data: Fouling factor 0.2 m²K/W; specific heat capacity of the coolant 4.18 kJ/kgK; heat transfer coefficient at the inside surface 850 W/m²K and heat transfer coefficient at the outside surface 600 W/m²K.

Solution
Use will be made of the tables of *Thermodynamic and Transport Properties of Fluids* for data relating to the refrigerant. Considering the primary fluid

$$Q = M_h(h - h_f) \text{ W}$$

From the tables the following data is obtained for refrigerant 134a: $h = 435.44$ kJ/kg, $h_f = 241.69$ kJ/kg. Thus at 7.7 bar absolute, condenser output

$$Q = 0.025(435.44 - 241.69) = 4.844 \text{ kW}$$

Now considering the secondary fluid, $Q = M_c \cdot C_c \cdot dt$. Substituting

$$4.844 = 0.08 \times 4.18(t_{c2} - 12)$$

from which

$$t_{c2} = 26.5°C$$

From the tables the following data is obtained for refrigerant 134a at 7.7 bar absolute: the superheat temperature is 50°C and the saturation temperature is 30°C.

$$\begin{array}{rcc}
\text{For counterflow: primary fluid} & 50 & \to 30 \\
\text{secondary fluid} & 26.5 & \leftarrow 12 \\
\hline
dt_{\text{max,min}} & 23.5 & 18
\end{array}$$

LMTD $dt_m = (23.5 - 18)/\ln(23.5/18) = 20.63$ K

The overall heat transfer coefficient

$$U = 1/R_t \text{ kW/m}^2\text{K} \quad \text{and} \quad R_t = (1/h_{si}) + R_f + (1/h_{so})$$
$$= (1/0.85) + 0.0002 + (1/0.6)$$
$$= 2.8433 \text{ m}^2\text{K/kW}$$

and therefore

$$U = 1/2.8433 = 0.3517 \text{ kW/m}^2 \text{ K}$$

Given

$$Q = U \cdot A \cdot dt_m \text{ kW}$$

Heating surface $A = Q/U \cdot dt_m = 4.844/0.3517 \times 20.63 = 0.668 \text{ m}^2$. The output of the condenser is calculated above as 4.844 kW.

9.4 Heat exchanger performance

The performance of heat exchangers with a solid boundary between the primary and secondary fluids depends upon the overall heat transfer coefficient U, which acts as the interface between the two fluids. This interface consists of three elements plus the fouling resistance:

- the hot side laminar sublayer,
- the solid interface or boundary,
- the cold side laminar sublayer.

The main sources of thermal resistance are the two laminar sublayers. In streamline or laminar flow the laminar sublayers offer appreciable thermal resistance because they have significant thickness through which heat must be conducted. Turbulent flow reduces this thickness and baffles are sometimes employed to increase turbulence. Turbulence can be induced in a fluid when $Re > 2000$. The overall heat transfer coefficient is therefore dependent upon fluid velocity on both sides of the solid boundary and upon the fouling resistance. The laminar sublayer in turbulent flow is considered in Chapter 6. The outside surface of the heat exchanger may be exposed to a fluid which has a lower specific heat capacity than that of the primary fluid. An example would be the case of an air heater battery supplied from a low temperature hot water heating system. The primary fluid is water having a specific heat capacity of about 4.2 kJ/kgK compared to that of air which has a specific heat capacity of around 1.0 kJ/kgK. In order to increase the heat transfer potential, extended finning is adopted on the air side of the battery. This increases the surface area to compensate for the lower specific heat capacity of air. The surface area of the heat exchanger should be a maximum within the limits of cost and size. Consider a calandria consisting of:

(a) 20 tubes with an inside diameter of 40 mm.
(b) 80 tubes with an inside diameter of 20 mm.

Both tube bundles will fit into the same size shell of 2.5 m in length. The respective surface areas are:

(a) $A = 20\pi \times 0.04 \times 2.5 = 6.283 \text{ m}^2$
(b) $A = 80\pi \times 0.02 \times 2.5 = 12.567 \text{ m}^2$

Clearly (b) surface is to be preferred.

The true temperature difference between fluids also has a direct influence upon the output of the calandria since it is the motive force in heat transfer. The minimum true temperature difference should not be less than 15–20 K for good heat exchange. There are a number of terms used in relation to heat exchangers which describe their performance and allow comparisons to be made. They include:

- capacity ratio (CR)
- effectiveness (E)
- number of transfer units (NTU).

Capacity ratio is the ratio of the products of mass flow and specific heat of each of the primary and secondary fluids. The product $M \cdot C$ is the thermal capacity of the moving fluid, the units of the terms being: $(\text{kg/s}) \times (\text{kJ/kgK}) = \text{kJ/sK} = \text{kW/K}$.

Capacity ratio is the ratio of the smaller product to that of the larger and therefore CR < 1.0.

If $(M_h \cdot C_h) > (M_c C_c)$, $\text{CR} = (M_c C_c)/(M_h C_h)$

If $(M_c \cdot C_c) > (M_h C_h)$, $\text{CR} = (M_h C_h)/(M_c C_c)$

where CR, being a ratio, is dimensionless.
There are two special cases to consider:

1. The capacity ratio becomes zero for both boiling and condensing where the units for C are kJ/kgK. If the evaporation or condensation of the fluid occurs at constant temperature, the temperature drop is 0 K.
2. The capacity ratio becomes unity (1.0) for equal thermal capacities of the primary and secondary fluids.

Effectiveness is the ratio of energy actually transferred to the maximum theoretically possible. Again it depends upon the product of mass flow and specific heat capacity of the primary and secondary fluids in kW/K.

If $(M_h \cdot C_h) > (M_c C_c)$, $E = (t_{c2} - t_{c1})/(t_{h1} - t_{c1})$

If $(M_c \cdot C_c) > (M_h C_h)$, $E = (t_{h1} - t_{h2})/(t_{h1} - t_{c1})$

where E, being a ratio, is dimensionless.

Regimes of fluid flow in heat exchangers 221

In parallel flow, t_{c2} approaches t_{h2} but can never exceed it whereas in counterflow t_{c2} can exceed t_{h2} and hence heat exchange in counterflow can be more effective. Refer to Figures 9.1 and 9.2.

Number of transfer units is the ratio of the product of the overall heat transfer coefficient and heat exchange area, and the thermal capacity $M \cdot C$ of either the primary or secondary fluid. The units of the product of the terms $U \cdot A$ are (kW/m²K) × m² = kW/K and since these are the same as the product of the terms $M \cdot C$, the number of transfer units, like CR and E, is a dimensionless quantity.

The ratio NTU was developed by W.M. Kays and A.L. London and published in 1964.

If $(M_h \cdot C_h) > (M_c C_c)$, NTU $= (U \cdot A)/(M_c \cdot C_c)$

If $(M_c \cdot C_c) > (M_h C_h)$, NTU $= (U \cdot A)/(M_h \cdot C_h)$

Capacity ratio, effectiveness and number of transfer units provide a straightforward route in the determination of the leaving temperatures of the primary and secondary fluids t_{h2} and t_{c2}, and in the heat exchanger output Q.

For counterflow heat exchangers,

$$E = [1 - \exp(-NTU(1 - CR))]/[1 - CR \cdot \exp(-NTU(1 - CR))]$$

When CR = 0, $E = [1 - \exp(-NTU)]$.
When CR = 1, $E = NTU/(NTU + 1)$.

For parallel heat exchangers,

$$E = [1 - \exp(-NTU(1 + CR))]/(1 + CR)$$

When CR = 0, $E = [1 - \exp(-NTU)]$.
When CR = 1, $E = [1 - \exp(-2NTU)]/2$.

There now follows some examples using the heat exchanger indices described above.

Example 9.5 Determine the effectiveness and fluid outlet temperatures of an economiser

(a) Determine the effectiveness and fluid outlet temperatures of an economiser handling 0.8 kg/s of flue gas at an inlet temperature of 280°C. The mean specific heat capacity is 1.02 kJ/kgK and boiler feed water entering at 0.6 kg/s and 60°C passes in parallel flow. The heat transfer surface is 1.8 m² and the overall heat transfer coefficient is known to be 1.85 kW/m²K. Take the mean specific heat capacity of the feed water as 4.24 kJ/kgK.
(b) What likely effect would a counterflow heater exchanger have on the flue gas?

222 Regimes of fluid flow in heat exchangers

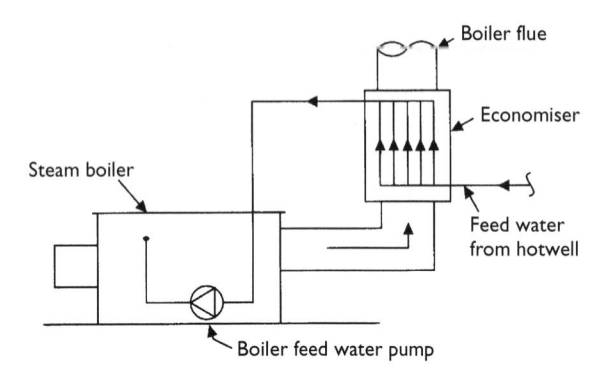

Figure 9.5 Location of the economiser for Example 9.5.

Solution (a)
Figure 9.5 shows a typical arrangement in which the economiser is used to extract heat from the boiler flue gases for heating the boiler feed water. By determining the NTU and the CR for the economiser, the effectiveness of its parallel flow heat exchanger can be evaluated.

First of all, however, the products of mass flow and specific heat capacity of the primary and secondary fluids must be calculated.

$$M_h \cdot C_h = 0.8 \times 1.02 = 0.816 \text{ kW/K}$$
$$M_c \cdot C_c = 0.6 \times 4.24 = 2.544 \text{ kW/K}$$

Therefore, since the capacity ratio is the smaller thermal capacity over the greater, CR = 0.816/2.544 = 0.32 and NTU = 1.85 × 1.8/0.816 = 4.081.

Effectiveness (E) can now be evaluated and

$$E = [1 - \exp(-4.081(1 + 0.32))]/(1 + 0.32)$$

from which

$$E = (1 - 0.00458)/1.32 = 0.754$$

Note: $\exp(-4.081(1 + 0.32)) = (2.7183)^{-5.3869} = 0.00458$
Since

$$E = (t_{h1} - t_{h2})/(t_{h1} - t_{c1})$$

then

$$0.754 = (280 - t_{h2})/(280 - 60)$$

from which

$$t_{h2} = 114°C$$

Using the heat balance: heat lost by flue gas = heat gained by feed water, that is,

$$0.816(280 - 114) = 2.544 \cdot dt$$

From which the temperature rise in the feed water $dt = 53$ K, and therefore $t_{c2} = 53 + 60 = 113°C$.

Solution (b)
For counterflow

$$E = [1 - \exp(-4.081(1 - 0.32))]/[1 - 0.32\exp(-4.081(1 - 0.32))]$$

from which

$$E = (1 - 0.0623)/(1 - 0.01995)$$

and

$$E = 0.957$$

This shows that counterflow is more effective than parallel flow where $E = 0.754$.

Since $M_c \cdot C_c > M_h \cdot C_h$,

$$E = (280 - t_{h2})/(280 - 60)$$

from which

$$t_{h2} = 70°C$$

Adopting the heat balance: heat lost by flue gas = heat gained by feed water, that is,

$$0.816(280 - 70) = 2.544 \cdot dt$$

from which the temperature rise in the boiler feed water $dt = 67$ K and the leaving temperature $t_{c2} = 67 + 60 = 127°C$.

SUMMARY FOR EXAMPLE 9.5
Clearly the counterflow heat exchanger is more effective. However, too much heat may be being extracted from the flue gas as t_{h2} at 70°C is likely to be below the dew point of the flue gas and corrosion would have to be accounted for in the chimney.

Example 9.6 Calculate the heat exchange surface, capacity ratio, effectiveness and number of transfer units for an HWS calorifier

Steam (0.18 kg/s) at 3.5 bar absolute, 0.9 dry enters a counterflow heat exchanger serving a hot water service (HWS) storage calorifier and condensate leaves at a temperature of 138.9°C. Feed water enters the calorifier at 10°C at the rate of 1.5 kg/s to satisfy the simultaneous HWS demand.

The heat transfer coefficients at the inside and outside surfaces of the heat exchanger are 13 kW/m²K and 10 kW/m²K respectively. Fouling resistance and the thermal resistance of the solid boundary between the primary and secondary fluids may be ignored. Given the specific heat capacity of water as 4.2 kJ/kgK, determine, for the heat exchanger in the HWS calorifier, its capacity ratio, effectiveness, number of transfer units and heat exchange surface.

Solution

The tables of *Thermodynamic and Transport Properties of Fluids* will be needed for the solution. From the tables it can be seen that there is no change in temperature of the primary steam which gives up its latent heat only in the heat exchanger. The larger thermal capacity (where for C here in kJ/kgK, temperature difference K being zero) is infinite. Therefore, the capacity ratio, $CR = 0$.

Using the heat balance to evaluate the leaving temperature of the secondary water t_{c2},

heat lost by the primary steam = heat gained by the secondary water

$$M_h(q \cdot h_{fg}) = M_c \cdot C_c \cdot dt$$

Note from the tables that the heat lost by the steam takes place at a constant temperature of 138.9°C. This, therefore, is a case of condensation at constant temperature. Refer to Figure 9.4.

Substituting from the tables and the data in the question:

$$0.18 \times (0.9 \times 2148) = 1.5 \times 4.2(t_{c2} - 10)$$

from which the secondary outlet temperature $t_{c2} = 65.23°C$.

Since the primary fluid is at constant temperature the arithmetic mean temperature difference between the fluids is the true mean value and

$$dt_m = 138.9 - (65.23 + 10)/2 = 101.29 \text{ K}$$

The overall heat transfer coefficient $U = 1/R_t$ kW/m²K
where

$$R_t = (1/h_{si}) + (1/h_{so}) = (1/13) + 1/10 = 0.1769 \text{ m}^2\text{K/W}$$

and

$$U = 1/0.1769 = 5.652 \text{ kW/m}^2\text{K}.$$

Now the rate of heat transfer at the heat exchanger can be obtained from either side of the heat balance, thus

$$Q = 0.18 \times (0.9 \times 2148) = 348 \text{ kW}$$

Since

$$Q = U \cdot A \cdot dt$$

heat exchanger surface

$$A = Q/U \cdot dt = 348/5.652 \times 101.29 = 0.608 \text{ m}^2$$

NTU $= U \cdot A/M_c \cdot C_c$ since the secondary water has the lower thermal capacity whereas the thermal capacity of the primary steam is zero. So NTU $= 5.652 \times 0.608/1.5 \times 4.2 = 0.545$.

Since capacity ratio CR $= 0$, the effectiveness of the heat exchanger

$$E = 1 - \exp(-\text{NTU}) = 1 - 2.7183^{(-0.545)} = 0.42$$

SUMMARY FOR EXAMPLE 9.6
Capacity ratio CR $= 0$, effectiveness $E = 0.42$, number of transfer units NTU $= 0.545$ and the heat exchange surface $A = 0.608$ m^2.

9.5 Cross flow

Figure 9.6 shows a typical air heater battery through which the primary fluid is constrained within the heat transfer tubes and over which air flows freely. The primary fluid is unmixed as it is contained within the tube boundary walls while the air which is the secondary fluid is considered as mixed flow. This requires the introduction of a correction factor f in the equation for the heat transfer across the heat exchanger. Thus $Q = U \cdot A \cdot f \cdot dt_m$ kW.

The LMTD dt_m is calculated for cross flow in the same way as for counterflow. Figure 9.7 shows how correction f can be evaluated for a cross flow heat exchanger with one fluid mixed and the other unmixed. When applying the factor it does not matter whether the hotter fluid is mixed or unmixed.

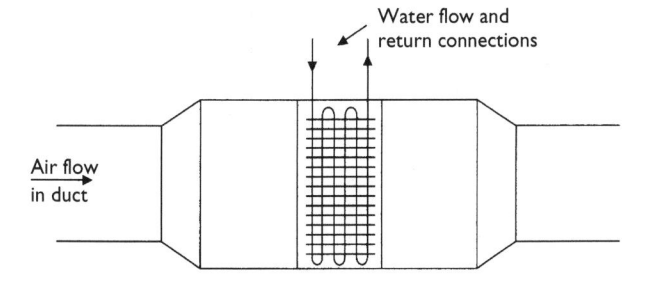

Figure 9.6 A cross flow air heater battery.

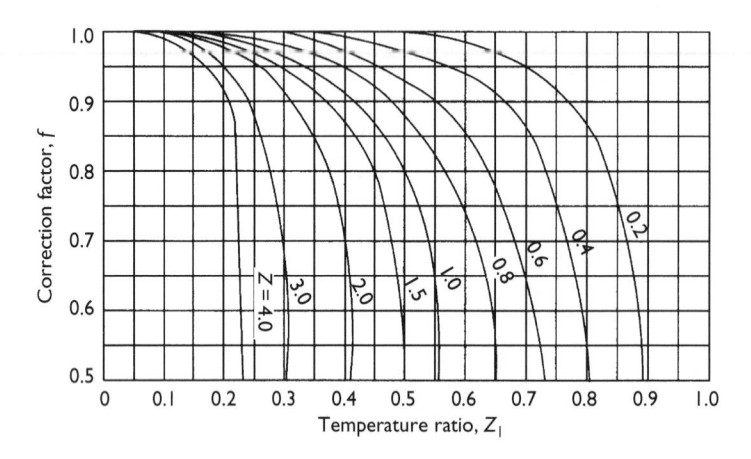

Figure 9.7 LMTD correction factor for cross flow, one fluid mixed, one fluid unmixed.

Example 9.7 Determine the heat exchange surface, capacity ratio, effectiveness and number of transfer units for a cross flow air heater battery

An air heater battery is supplied with water at 1.0 kg/s and 82°C with return water at 72°C. Air enters the heater at 1.86 m³/s and 20°C and leaves at 50°C.

(a) Assuming cross flow, determine correction factor f from Figure 9.7 and hence calculate the heat exchanger surface.
(b) Determine the capacity ratio, effectiveness and the number of transfer units for the heater battery.

Data: Specific heat capacities for water and air are 4.2 and 1.0 kJ/kgK, respectively; the heat transfer coefficients at the inside and outside surface of the exchanger tubes are 3.72 and 2.0 kW/m² K, fouling resistance is 0.0002 m² K/kW. Air density is 0.9 kg/m³.

Solution (a)
The horizontal axis of Figure 9.7,

$$Z_1 = (t_{h2} - t_{h1})/(t_{c1} - t_{h1}) = (70 - 82)/(20 - 82)$$

Thus $Z_1 = (-12)/(-62)$, from which the temperature ratio $Z_1 = 0.19$. From the same figure,

$$Z = (t_{c1} - t_{c2})/(t_{h2} - t_{h1}) = (20 - 50)/(70 - 82) = (-30)/(-12)$$

from which $Z = 2.5$. Adopting the values for temperature ratios Z and Z_1, the correction factor from Figure 9.7 is $f = 0.96$.

The overall heat transfer coefficient $U = 1/R_t$ kW and

$$R_t = (1/3.72) + 0.0002 + (1/2) = 0.769 \text{m}^2 \text{ K/kW}$$

then

$$U = 1/0.769 = 1.3 \text{ kW/m}^2 \text{ K}$$

Taking cross flow as counterflow to obtain LMTD without loss of integrity:

primary fluid:	water	82 → 70
secondary fluid:	air	50 ← 20
$dt_{min,max}$		32 50

LMTD $dt_m = (50 - 32)/\ln(50/32) = 40.3$ K

from

$$Q = U \cdot A \cdot dt_m \cdot f \quad A = Q/U \cdot dt_m \cdot f$$

where

$$Q = M_h \cdot C_h \cdot dt = 1.0 \times 4.2 \times (82 - 70) = 50.4 \text{ kW}$$

and substituting, heating surface

$$A = 50.4/1.3 \times 40.3 \times 0.96 = 1.0021 \text{ m}^2$$

Solution (b)

Now

$$M_h \cdot C_h = 1.0 \times 4.2 = 4.2 \text{ kW/K}$$

and

$$M_c \cdot C_c = (1.86 \times 0.9) \times 1.0 = 1.674 \text{ kW/K}$$

The capacity ratio CR = 1.674/4.2 = 0.399
Number of transfer units NTU = $U \cdot A/M_c \cdot C_c$ = 1.3 × 1.0021/1.674 = 0.778.

The determination of effectiveness for a cross flow heat exchanger, one fluid mixed, one fluid unmixed, is derived from the relationship between capacity ratio and number of transfer units. An approximate value for effectiveness here may be calculated assuming counterflow.

Approximate effectiveness, taking cross flow as counterflow,

$$E = [1 - \exp(-0.778(1 - 0.399))]/[1 - 0.399 \exp(-0.788(1 - 0.399))]$$
$$= (1 - 0.6265)/(1 - 0.25) = 0.498$$

The actual answer is $E = 0.484$

228 Regimes of fluid flow in heat exchangers

SUMMARY FOR EXAMPLE 9.7

(a) Heating surface $A = 1.0021$ m^2.
(b) CR $= 0.399$, $E = 0.498$, NTU $= 0.778$.

The solution must be qualified to the extent that the true temperature difference between fluids and the effectiveness was determined for counterflow.

9.6 Further examples

Example 9.8 Calculate the mass transfer of the secondary fluid and the length of the primary tube bundle for a non-storage heating calorifier

A shell and tube non-storage heating calorifier operates in counterflow. The primary medium is high temperature hot water at temperatures of 160°C and 130°C. The secondary medium is low temperature hot water operating at 82°C and 70°C.

The heat exchange tube bundle consists of four copper tubes each with 20 mm inside diameter and 25 mm outside diameter having a thermal conductivity of 0.35 kW/mK and heat transfer coefficients of 5 and 3 kW/m^2 K, respectively.

Ignoring the effects of fouling on the heat exchange surfaces, determine the mass flow of the secondary medium and the length of the tube bundle. Take the specific heat capacities of the primary and secondary mediums at the appropriate mean water temperature and the mass flow of high temperature hot water as 0.349 kg/s.

Solution
The shell and tube heat exchanger is shown in Figure 9.8.

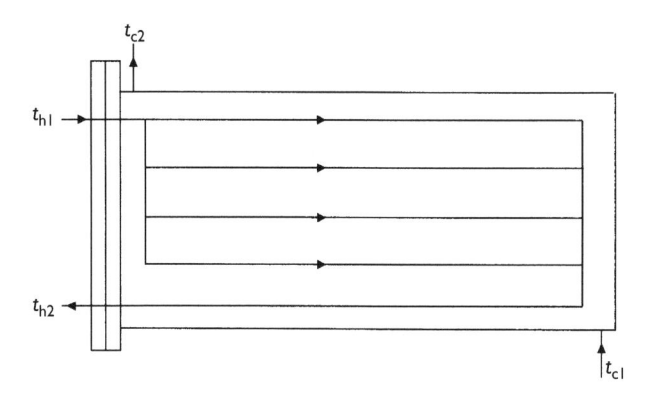

Figure 9.8 Non-storage calorifier (Example 9.8).

Regimes of fluid flow in heat exchangers 229

The mean water temperature of the primary medium $= (160 + 130)/2 = 145°C$. The mean water temperature of the secondary medium $= (82 + 70)/2 = 76°C$. From the tables of *Thermodynamic and Transport Properties of Fluids* the specific heat capacities at these mean water temperatures are: primary medium, 4.3 kJ/kgK and secondary medium, 4.194 kJ/kgK. Adopting the heat balance:

heat lost by the primary medium = heat gain by the secondary medium

and substituting:

$$0.349 \times 4.3 \times (160 - 130) = M \times 4.194 \times (82 - 70)$$

from which the mass transfer of low temperature hot water is 0.895 kg/s.

For counterflow, primary fluid: 160 → 130
secondary fluid: 82 ← 70
$dt_{max,min}$ 78   60

LMTD $dt_m = (78 - 60)/\ln(78/60) = 68.6$ K.

Since the surface of the heat exchanger is identified as four copper tubes it is convenient to determine the overall heat transfer coefficient for radial conductive heat flow which is introduced in Chapter 2.

Thus from Chapter 2, equation 2.12:

$$Q/L = (2\pi\, dt)/[(1/r_1 \cdot h_{si}) + ((\ln r_2/r_1)/k_1) + (1/r_2 \cdot h_{so})] \text{ W/m run}$$

Since

$$Q/L = U_L \cdot dt \text{ W/m}$$

therefore

$$U_L = (Q/L)/dt \text{ W/mK}$$

then, as dt cancels

$$U_L = (2\pi)/[(1/r_1 \cdot h_{si}) + ((\ln(r_2/r_1)/k_1) + (1/r_2 \cdot h_{so}) \text{ W/mK}.$$

Note the different units for the overall heat transfer coefficient which is in per metre run of pipe and not per square metre of plane surface (see summary below).
Substituting:

$$U_L = (2\pi)/[(1/0.01 \times 5) + (\ln(0.0125/0.01)/0.35) + (1/0.0125 \times 3)]$$

230 Regimes of fluid flow in heat exchangers

from which

$$U_L = (2\pi)/[20 + 0.6376 + 26.667]$$

thus

$$U_L = 0.133 \text{ kW/mK}$$

The output of the heat exchanger can be obtained from either side of the heat balance:
thus

$$Q = 0.349 \times 4.3 \times (160 - 130) = 45 \text{ kW}$$

now

$$Q/L = U_L \cdot dt_m \text{ kW/m}$$

and therefore

$$Q = U_L \cdot L \cdot dt_m \text{ kW}$$

Thus for a four tube bundle:
Length

$$L = (Q/U_L \cdot dt_m) \times 1/4 = [45/(0.133 \times 68.6)] \times 1/4 = 1.233 \text{ m}$$

SUMMARY FOR EXAMPLE 9.8
The mass flow of low temperature hot water is 0.85 kg/s and the length of the tube bundle is 1.233 m which does not account for the inefficiency of heat exchange. Note the low thermal resistance of the copper tubes relative to the heat transfer coefficients at the inner and outer surfaces. This is the reason why the thermal resistance of the solid boundary is often ignored. If mild steel heat exchange tubes were used, the thermal conductivity of the material would be in the region of 0.05 kW/mK and the thermal resistance of the solid boundary would be:

$$R = [(\ln(0.0125/0.01)/(0.05)] = 4.463 \text{ m}^2 \text{ K/kW}.$$

This compares with that for copper tubes of $R = [\ln(0.0125/0.01)/(0.35)] = 0.6376 \text{ m}^2 \text{ K/kW}$ which is much reduced. You can see why copper is favoured as the heat exchange material.

The overall heat transfer coefficient (U) in kW/m^2 K can be calculated:
The outside diameter of the copper tubes is 25 mm and for 1 m run the surface area will be $A = \pi \cdot d \cdot L = \pi \times 0.025 \times 1.0 = 0.07854 \text{ m}^2$
So $U = U_L/A = 0.133/0.07854 = 1.6934 \text{ kW/m}^2 \text{ K}$

Example 9.9 Determine the condenser output, the leaving temperature of the coolant and the length of the tube bundle for a vapour compression refrigration unit

Tetrafluoroethane is discharged from a compressor at 14.91 bar absolute having 20 K of superheat and enters a condenser at 0.3 kg/s. It leaves the condenser subcooled by 5 K. The coolant flow rate is 0.794 kg/s and the inlet temperature is 16°C. Assuming counterflow determine the condenser output, the leaving temperature of the coolant and the length of the tube bundle.

Data: The tube bundle consists of eight 15 mm nominal bore tubes having an outer diameter of 20 mm, the overall heat transfer coefficient is 3.2 kW/m² K and the specific heat capacity of the coolant is 2.8 kJ/kgK.

Solution
Reference should be made to the tables of *Thermodynamic and Transport Properties of Fluids*. From the tables refrigerant 134a at 14.91 bar absolute has a specific enthalpy of 449.45 kJ/kgK at 20° of superheat. The temperature of the superheated vapour is 75°C and at saturated conditions it is 55°C. Thus since it is subcooled by 5 K on leaving the condenser the leaving temperature of the refrigerant will be 55 − 5 = 50°C. The specific enthalpy of the refrigerant leaving the condenser will therefore be 271.61 kJ/kgK. You should now confirm these data from the tables.
The condenser output

$$Q = M(h - h_f) \text{ kW}$$

Substituting:

$$Q = 0.3(449.45 - 271.61) = 53.35 \text{ kW}$$

Adopting the heat balance:

heat lost by refrigerant = heat gained by coolant

Substituting:

$$53.35 = 0.794 \times 2.8 \text{ d}t$$

from which the temperature rise of the coolant dt = 24 K
therefore the leaving temperature of the coolant $t_{c2} = 16 + 24 = 40°C$

For counterflow, primary fluid   75 → 50
                     secondary fluid   40 ← 16
                           d$t_{max,min}$ 35   34

LMTD d$t_m = (35 - 34)/\ln(35/34) = 34.5$ K

Note that since the maximum and minimum temperature differences are almost equal the arithmetic mean temperature difference can be taken and equals $(62.5 - 28) = 34.5$ K. For the determination of tube bundle length $Q = U_L \cdot L \cdot dt_m$ kW, where U_L has the units kW/m·runK.

Now the overall heat transfer coefficient

$$U = 3.2 \text{ kW/m}^2 \text{ K}$$

and

$$Q = U \cdot A \cdot dt_m = 53.35 \text{ kW}$$

Then total area of tube bundle

$$A = 53.35/(3.2 \times 34.5) = 0.4832 \text{ m}^2$$

Also

$$A = \pi \cdot d \cdot L$$

so

$$L = A/(\pi \times d) = 0.4832/(\pi \times 0.02) = 7.69 \text{ m}$$

The eight tube bundle length

$$L = 7.69/8 = 0.961 \text{ m}$$

SUMMARY FOR EXAMPLE 9.9

Condenser output 53.35 kW, leaving temperature of the coolant 40°C, length of the tube bundle 0.961 m.

The heat exchanger will be similar to Figure 9.8 but would have eight tubes in the bundle instead of the four shown in the diagram.

9.7 Chapter closure

This completes the work on heat exchangers, only a few of which have been considered in detail here. However, the principles of heat exchanger design and performance have been introduced from which you will appreciate that the subject is very specialised and largely in the domain of the manufacturer. It is important though for the student in building services to have some knowledge of this work and you should now extend it by undertaking market research into the types of heat exchangers available and their applications.

Appendix 1

Verifying the form of an equation by dimensional analysis

Nomenclature

A	area (m²)
C	specific heat capacity (kJ/kgK)
d	diameter (m)
dt	temperature difference (K)
f	frictional coefficient
g	gravitational acceleration (m/s²)
h	head (m)
H	dimension for energy (J, Nm, Ws, kWh)
I	flux density (W/m²)
L	length (m)
L	dimension for length (m)
LHS	left hand side
M	mass flow (kg/s)
M	dimension for mass (kg)
P	pressure (Pa)
Q	rate of heat flow (W), volume flow (m³/s)
RHS	right hand side
T	dimension for time (s)
TE	total energy; metres of fluid flowing
U	thermal transmittance coefficient (W/m²K)
u	velocity (m/s)
ρ	density (kg/m³)
σ	Stefan–Boltzman constant (W/m²K⁴)
μ	absolute viscosity (kg/ms)
θ	dimension for thermodynamic temperature (K)

A1.1 Introduction

The process of checking the units of the terms in an equation is relatively straightforward and a common strategy for ensuring that its form is correct. For example, the formula $Q = U \cdot A \cdot dt$ has the units of Watts. This unit for the term Q can be checked if the units of the other terms are known, thus the product of the terms $U \cdot A \cdot dt$ has the units $(W/m^2 K) \times (m^2) \times (K)$ and by the process of cancellation the unit of Watts (W) is confirmed.

Dimensions, on the other hand, are properties (of a term) which can be measured. For example, density has the units kg/m^3. Its dimensions are ML^{-3} where M is mass and L is length. Units are the elements by which numerical values of these dimensions describe the term quantitavely. That is to say, the units of density are kg/m^3 where mass is quantified in kg and volume is quantified in m^3. Thus the units of a term define, in addition, the system of measurement being used.

A term's dimensions, on the other hand, are not confined to any system of measurement and therefore dimensional analysis is universal and common to all systems of units. There are three systems of units in use in the Western world namely the System International (SI), FPS or foot pound second system and MKS or metre kilogram second system. The United Kingdom has for many years adopted the SI. The United States of America is slowly changing from the FPS system to SI and likewise with Germany from the MKS system of measurement.

Dimensional analysis is adopted to undertake three discrete tasks which are

- to check that an equation has been correctly formed
- to establish the form of an equation relating to a number of variables
- to assist in the analysis of empirical formulae in experimental work.

This chapter will focus on checking some equations used in this book by dimensional analysis to show that they are correctly formed.

AI.2 Dimensions in use

It is necessary to identify the *dimensions* which will be used in validating a formula. There are five *dimensions* which are used in heat and mass transfer, and they are those for mass M, length L, time T, energy H and thermodynamic temperature θ.

The *dimensions* for force, for example, can now be established and since force = mass × acceleration, the units of the terms are: force = $kg \times m/s^2$. Therefore, the *dimensions* of the terms are: force = $M \times L \times T^{-2} = MLT^{-2}$. The unit of force in the SI is the Newton (N), thus the *dimensions* of the Newton are MLT^{-2}.

The unit of the *dimension* for heat energy H is the Joule. This is the same as the units for mechanical energy which are the (Nm). So the *dimensions* for energy H are $(MLT^{-2}) \times (L)$. So the *dimensions* for $H = ML^2T^{-2}$.

The Newton and the Joule are called derived units since they are made up from more than one basic unit. There are six basic units in the SI and the *dimensions* of four of them, M, L, T and θ, will be used as well as

Table A1.1 Dimensions of some derived units used in heat and mass transfer

Term	Name	Symbol	Units, definitions	Dimensions
Force	Newton		kg × m/s²	MLT^{-2}
Energy	Joule (Nm)	H	force × distance moved	$ML^2 \cdot T^{-2}$
Power	Watt	P_w, Q	energy/time	$ML^2 \cdot T^{-3}$
Pressure	Pascal	P	force/area	$ML^{-1} \cdot T^{-2}$
Gravitational acceleration		g	m/s²	LT^{-2}
Density		ρ	kg/m³	ML^{-3}
Mass flow		M	kg/s	ML^{-1}
Volume flow		Q	m³/s	$L^3 \cdot T^{-1}$
Absolute viscosity		μ	kg/m·s	$ML^{-1} \cdot T^{-1}$
Kinematic viscosity		υ	absolute viscosity/density	$L^2 \cdot T$
Specific heat capacity		C	J/kgK	$L^2 \cdot T^{-2} \cdot \theta^{-1}$
Specific enthalpy		h	kJ/kg	$L^2 \cdot T^{-2}$
Cubical expansion		β	$1/\theta$	θ^{-1}
Thermal conductivity		k	W/mK	$MLT^{-3} \cdot \theta^{-1}$
Heat flux		l	W/m²	MT^{-3}
Speed of rotation		N	rev/s	T^{-1}
Heat transfer coefficient		h	W/m²K	$MT^{-3} \cdot \theta^{-1}$
Mean velocity		u	m/s	LT^{-1}
Area		A	m²	L^2

dimension H which is the derived unit for energy. These will be used to define the *dimensions* of terms in this appendix and Appendix 2. You will see the importance of knowing the basic units in a derived unit such as the Newton before identifying its *dimensions*. The derived units of terms commonly used in heat and mass transfer are listed in Table A1.1 and you should confirm the units and *dimensions* of each one before proceeding.

There now follows some examples in which the forms of some equations used in the book are checked by dimensional analysis.

Example A1.1 Verify the equation for pressure

$P = h \cdot \rho \cdot g$ Pa.

Solution
Referring to Table A1.1. The right-hand side (RHS) of the equation has the dimensions $(L)(ML^{-3})(LT^{-2})$. These reduce to $ML^{-1} \cdot T^{-2}$ which agrees with the dimensions for pressure in Table A1.1.

Example A1.2 Verify the D'Arcy equation

$$h = (4fLu^2)/(2gd) \text{ metres of fluid flowing.}$$

Solution
Referring to Table A1.1. The RHS of the formula has the dimensions $[(L)(L^2/T^2)/(L/T^2)(L)]$. Note that the pure numbers do not have dimensions. It is assumed for the moment that the coefficient of friction f is dimensionless. The dimensions on the RHS of the formula reduce to L. This shows that the form of the D'Arcy formula is correct and confirms that the coefficient of friction must be dimensionless.

Example A1.3 Show that the Reynolds number is dimensionless.

Show by dimensional analysis that the Reynolds number is dimensionless.

Solution
Now $Re = (\rho \cdot u \cdot d)/\mu$. Referring to Table A1.1 and substituting the dimensions on the RHS of the equation:

$$(ML^{-3})(LT^{-1})(L)/(ML^{-1} \cdot T^{-1}) = (ML^{-3})(LT^{-1})(L)(M^{-1}LT)$$

from which all the dimensions on the RHS cancel and therefore the Reynolds number is a dimensionless number.

Example A1.4 Verify Box's formula

Check the form of Box's formula for head loss in turbulent flow in straight pipes where:

$$h = (fLQ^2)/(3d^5) \text{ metres of fluid flowing.}$$

Solution
Referring to Table A1.1 and substituting the dimensions on the RHS of the formula:

$$(L)(L^3 \cdot T^{-1})^2(L^{-5}) = (L)(L^6 \cdot T^{-2})(L^{-5})$$

The RHS reduces to: $(L^2)(T^{-2})$

The RHS should reduce to dimension (L) to equate with the left-hand side (LHS) of the formula. Further cancellation leaves the dimensions $(L)(T^{-2})$ on the RHS and from Table A1.1 the term which will cancel these dimensions is the inverse of gravitational acceleration which is $(T^2)(L^{-1})$. You should now confirm that this is so. The conclusion therefore is that the constant (1/3) in the formula must include the term gravitational acceleration in its denominator.

In the derivation of Box's formula, (Chapter 6, equation 6.8) $h = (64/2\pi^2 \cdot g)(fLQ^2/d^5)$ metres of fluid flowing and gravitational acceleration (g) appears in the denominator of the constant which is enclosed by the first set of brackets. If (g) is taken as 9.81 m/s² the constant evaluates to 1/3. You should now confirm this is so. The analysis of dimensions therefore has shown that the form of Box's formula is correct. It also identifies the fact that the constant in the formula of 1/3 does have dimensions which are $(T^2)(L^{-1})$.

Example A1.5 Verify the mass transfer equation

Show that the equation for the determination of mass flow of water from the equation $Q = M \cdot C \cdot dt$, is formed correctly.

Solution
From Table A1.1, the dimensions of Q which is the rate of heat flow in Watts are: $ML^2 \cdot T^{-3}$. The dimensions of the terms on the RHS of the equation are, where specific heat capacity from Table A1.1 has the dimensions $(L^2 \cdot T^{-2} \cdot \theta^{-1})$: $(MT^{-1})(L^2 \cdot T^{-2} \cdot \theta^{-1})(\theta) = (ML^2 \cdot T^{-3})$ which agrees with the LHS of the equation and therefore the form of the equation is correct.

Example A1.6 Verify the Bernoulli equation

Show that the formula $TE = Z + (P/\rho g) + (u^2/2g)$ metres of fluid flowing is formed correctly.

Solution
This is the Bernoulli equation for the conservation of energy. Referring to Table A1.1.

Total Energy $TE = (L) + (ML^{-1} \cdot T^{-2})/(ML^{-3})(LT^{-2}) + (L^2T^{-2})/(LT^{-2})$

from which $TE = (L) + (L) + (L)$ and the formula for total energy is shown to be formed correctly.

Example A1.7 Verify the dimensions for the Stefan–Boltzman constant for heat radiation

Adopting dimensional analysis, verify the dimensions of the Stefan–Boltzman constant for heat radiation.

Solution
The equation is flux density $I = \sigma \cdot T^4$ W/m². Referring to Table A1.1, the dimensions of heat flux on the LHS of the equation are: MT^{-3}. The RHS

of the formula is the product of a numerical constant and thermodynamic temperature to the fourth power thus: $\sigma \cdot T^4$.

In order for the RHS of the formula to be dimensionally similar to the LHS the product of the Stefan–Boltzman constant and thermodynamic temperature to the fourth power must have the dimensions: $(MT^{-3} \cdot \theta^{-4})(\theta^4)$. This will then reduce the RHS of the formula to MT^{-3} which is dimensionally similar to the LHS.

Thus the dimensions of the Stefan-Boltzman constant will be: $MT^{-3} \cdot \theta^{-4}$.

You will note that not all numerical constants in formulae are dimensionless. Example A1.4 is another case in point.

AI.3 Appendix closure

You have now been introduced to the *dimensions* of terms in common use and undertaken one of the applications to which dimensional analysis can be put. Clearly it is easier to check the form of an equation by using the units of the terms within it as shown in the Introduction A1.1, rather than by using the dimensions of the terms. However, it is important to persevere with using the dimensions of terms since they are used exclusively in Appendix 2 and they cross the boundaries of different systems of measurement.

This chapter has sought to illustrate the way the dimensions of terms are expressed and provides a grounding for Appendix 2, which is where dimensional analysis has a major role in establishing the form of an equation relating to a number of variables and in assisting the analysis of experimental work.

Appendix 2

Solving problems by dimensional analysis

Nomenclature

Some of the nomenclature used in this chapter are given in Table A1.1 of Appendix 1 to which reference will need to be made. The remainder is listed here.

- B constant of proportionality
- C specific heat capacity (J/kgK)
- d diameter, distance from leading edge (m)
- dh difference in head, metres of fluid flowing
- dP pressure drop (Pa)
- dt temperature difference (K)
- $d\theta$ thermodynamic temperature difference (K)
- f frictional coefficient for turbulent flow
- Gr Grashof number
- k thermal conductivity (W/mK)
- L length (m)
- m number of dimensions
- n number of variables
- Nu Nusselt number
- P pressure (Pa)
- Pi Buckingham's Pie theorem
- Pr Prandtl number
- Re Reynolds number
- u fluid velocity (m/s)
- β cubical expansion (K^{-1}) from $m^3/m^3 K$
- ϕ function of
- τ shear stress (Pa)
- θ thermodynamic temperature (K)

A2.1 Introduction

Appendix 1 introduces one of the three applications of dimensional analysis. The two remaining applications include: establishing the form of an equation relating a number of variables and assisting in establishing empirical formulae in experimental work. The dimensions which will be used in this chapter continue to be those for mass M, length L, time T and thermodynamic temperature θ. The dimension for heat energy H is used in Example A2.7

A2.2 Establishing the form of an equation

A formula which is well known to building services will be used now to demonstrate this use of dimensional analysis.

Case study A2.1 Verify the form of the D'Arcy equation

Consider the D'Arcy equation for turbulent flow which may be written as: $dh = (4fLu^2)/2gd$ metres of fluid flowing. This formula may be in terms of units of pressure and since $dP = dh \cdot \rho \cdot g$ Pa, the D'Arcy equation can be expressed as $dP = [(4fLu^2)/2gd]\rho \cdot g$ Pa, thus $dP = (4fLu^2\rho)/2d$ Pa. If it is required to express the formula in terms of a rate of pressure loss per metre: then

$$dP/L = (4fu^2\rho)/2d \text{ Pa/m} \tag{A2.1a}$$

Solution

We will now use dimensional analysis to find the form of an equation for turbulent flow using a number of variables to see whether it agrees with D'Arcy's formula A2.1a. With turbulent flow of a fluid in a flooded pipe, pressure loss per unit length is likely to be related to the variables u, ρ, μ and a characteristic dimension of the pipe or duct, say diameter d.

Note: Selecting the variables in a blind analysis may require iteration to find the appropriate variables for the task.

Thus

$$dP/L = \phi(u, \rho, \mu, d)$$

The term ϕ means 'a function of'. If an exponential relationship is assumed

$$dP/L = B[u^a \rho^b \mu^c d^d] \tag{A2.1b}$$

where B is taken as a constant of proportionality and the indices a, b, c and d may need to be evaluated empirically. The expression is now changed to the dimensions of the terms within it and using Table A1.1 to assist in defining the terms:

$$(ML^{-2} \cdot T^{-2}) = B[(LT^{-1})^a (ML^{-3})^b (ML^{-1} \cdot T^{-1})^c (L)^d]$$

Now forming equations from the indices and remembering, for example, that the dimension M is in fact M^1.

For dimension M:

$$1 = b + c$$

For dimension L:

$$-2 = a - 3b - c + d$$

For dimension T:

$$-2 = -a - c$$

From which

$$b = 1 - c, \; a = 2 - c \quad \text{and} \quad d = -1 - c.$$

The simplification of the indicial equation for dimension L needs clarification as follows:

For above dimension L:

$$-2 = a - 3b - c + d$$

Substitute $b = 1 - c$ and $a = 2 - c$
Then

$$-2 = (2 - c) - 3(1 - c) - c + d$$

Removing brackets

$$-2 = 2 - c - 3 + 3c - c + d$$

Simplifying

$$-2 = -1 + c + d$$

Rearranging in terms of d,

$$d = -1 - c$$

Substituting the expressions for a, c and d into equation A2.1b

$$dP/L = B[u^{(2-c)} \cdot \rho^{(1-c)} \cdot \mu^c \cdot d^{(-1-c)}]$$

Inspecting this statement identifies from the indices that there are two groups of variables on the right-hand side (RHS) of the equation namely one with the numerical index and one with the index $-c$. Putting the terms into these two groupings we have:

$$dP/L = B[(u^2\rho/d)(\rho u d/\mu)^{-c}]$$

You will notice that the last group of terms $(\rho u d/\mu)^{-c}$ is in fact the Reynolds number Re and it is found by experiment that the constant of proportionality $B=$ unity. Thus we have $dP/L = (u^2\rho/d)\phi(Re)$. The term ϕ accounting for the index $-c$.

Now the D'Arcy equation may be expressed in the form:

$$dP/L = 4fLu^2\rho/2d = (u^2\rho/d)(4f/2)$$

from which, therefore, $\phi(Re) = 4f/2$. The frictional coefficient f is obtained from the Moody diagram which has a scale of Reynolds numbers as one of its axes. Refer to Chapter 6, Figure 6.14.

SUMMARY FOR CASE STUDY A2.1
It is evident from the result of the analysis of the dimensions of the selected variables that the equation so derived is comparable with the D'Arcy formula for turbulent flow. It is likely that D'Arcy used dimensional analysis to arrive at his formula which left him to sort out the constant of proportionality B and $\phi(Re)$ by experiment. Dimensional analysis had reduced his investigation by finding the form of the equation.

Buckingham's Pi theorem

This states that if there are n variables in a problem with m dimensions there will be $(n-m)$ dimensionless groups in the solution. Applying this theorem to Case study A2.1, from equation A2.1b there are $n=5$ variables. The term (dP/L) which is the subject of the expression is taken as one variable, the others are u, ρ, μ and d. Subsequently, three dimensions (M, L, T) were used in the analysis. Thus $(n-m) = (5-3) = 2$ dimensionless groups. The Reynolds number has been identified as one of the groups. The other group can now be found from: $dP/L = (u^2\rho/d)\phi(Re)$.

Then $\phi(Re) = (dP/L)(d/u^2\rho)$ and the second dimensionless group must be $(dP \cdot d/u^2\rho L)$, in order to keep the equation in balance. You should now check the dimensions of these terms to establish that this group is dimensionless.

A2.3 Dimensional analysis in experimental work

Where full scale experiments cannot be conducted because of the problem of size, information can be obtained by experiments on models, provided that the model is related properly to the full size counterpart. The relationship is simple if fluid flow, for example, is geometrically and dynamically similar in each case.

Geometrical similarity

This is achieved when one system is the scale model of the other; that is, the ratio of corresponding lengths is constant.

Dynamical similarity

This is achieved when several forces acting on corresponding fluid elements have the same ratio to one another in both systems. For example, in turbulent flow in flooded pipes, the value of the Reynoldss number must be identical in the scale model and its full size counterpart. In Case study A2.1, both dimensionless groups would need to satisfy this condition if a model of a full size version of a system involving turbulent flow in horizontal straight pipes was being built. Thus,

$$(Re)_m = (Re)_{fs} \quad \text{and} \quad (dP \cdot d/u^2 \rho L)_m = (dP \cdot d/u^2 \rho L)_{fs}$$

where subscript m refers to the model and subscript fs refers to the full size version. These dimensionless groups will now be used in the case study which follows.

Case study A2.2 Using a model to verify the pressure loss in a full size counterpart

The pressure loss in a pipe 50 mm bore and 10 m long (the model) is found to be 15 kPa when water flows full bore at a mean velocity of 2.5 m/s. Determine the pressure loss when sludge flows full bore at the corresponding speed through a pipe 300 mm bore and 360 m long (full size counterpart).

Data: Density of water and sludge is 1000 and 1200 kg/m^3 respectively, absolute viscosity for water and sludge is 0.0013 and 0.002 kg/ms respectively.

Solution
For water the Reynolds number

$$Re = \rho u d/\mu = (1000 \times 2.5 \times 0.5)/0.0013 = 96\,154$$

Flow is therefore fully turbulent and since this is the case, the formula appropriate to this solution is the one analysed in Case study A2.1 $(dP \cdot d/u^2 \rho L) = \phi(Re)$. For geometric similarity the model must be of a similar scale to the full size counterpart. Since flow is full bore, the area ratio of the pipe cross section is $(\pi d_m^2/4)/(\pi d_{fs}^2/4)$. The ratio simplifies to $d_m^2/d_{fs}^2 = (50)^2/(300)^2 = 0.0278$. The length to scale of the pipe in the model must be $0.0278 \times 360 = 10$ m and this is the case.

For dynamic similarity

$$(Re)_m = (Re)_{fs} \quad \text{and} \quad (dP \cdot d/u^2 \rho L)_m = (dP \cdot d/u^2 \rho L)_{fs}$$

For equality of Reynolds numbers $Re = \rho u d/\mu$. Substituting:

$$(1000 \times 2.5 \times 0.05/0.0013)_m = (1200 \times u \times 0.3/0.002)_{fs}$$

from which the corresponding mean velocity of the sludge $u = 0.534$ m/s.

For equality of the second dimensionless group $(dP \cdot d/u^2 \rho L)$, substituting:

$$\{15 \times 0.05/[(2.5)^2 \times 1000 \times 10]\}_m$$
$$= \{dP \times 0.3/[(0.534)^2 \times 1200 \times 360]\}_{fs}$$

from which the corresponding pressure drop $dP = 4.93$ kPa.

SUMMARY FOR CASE STUDY A2.2

By employing geometrical and dynamical similarity we have correctly estimated the pressure drop in the pipe carrying the sludge by using the scale model. Now that we have calculated the corresponding velocity of the sludge from equating the Reynolds number, the D'Arcy formula can now be used to check the pressure drop along the pipe carrying the sludge, where $dP = 4fLu^2\rho/2d$ Pa. Substituting

$$dP = [4 \times 0.006 \times 360 \times (0.534)^2 \times 1200]/(2 \times 0.3) = 4927 \text{ Pa}$$
$$= 4.93 \text{ kPa}$$

which agrees with the solution above.

Using D'Arcy's formula you should now confirm that the pressure drop in the model is 15 kPa. Note that the frictional coefficient for turbulent flow was taken as $f = 0.006$ for both the model and the sludge pipe.

A2.4 Examples in dimensional analysis

Example A2.1 Verify the centrifugal pump and fan laws

The laws for centrifugal pumps and fans are: $Q \propto N$, $P \propto N^2$ and $P_w \propto N^3$. Verify these laws using dimensional analysis and identify the dimensions of the numerical constant B.

Solution
Let

$$Q = \phi(N) \tag{A2.2}$$

then $Q = B(N^a)$ where B and a are numerical constants. From Table A1.1, the dimensions of the terms are: $(L^3T^{-1}) = (T^{-1})$. Then

$$(L^3 \cdot T^{-1}) = T^{-a}$$

For L, $3 = 0$;
for T, $-1 = -a$, therefore $a = 1$.

Substituting into equation A2.2:

$$Q = \phi(N) = B(N)$$

If B is the constant of proportionality, $Q \propto N$. Identifying the dimensions of the constant B from $Q = B(N)$. Thus

$$(L^3 \cdot T^{-1}) = B(T^{-1})$$

From which the dimensions of the numerical constant $B = L^3$.
Let

$$P = \phi(N) \tag{A2.3}$$

then $P = B(N^a)$. From Table A1.1, the dimensions of the terms are: $(ML^{-1} \cdot T^{-2}) = (T^{-1})$. Then

$$(ML^{-1} \cdot T^{-2}) = T^{-a}$$

For M, $1 = 0$;
for L, $-1 = 0$;
for T, $-2 = -a$, from which $a = 2$.

Substituting into equation A2.3:

$$P = \phi(N^2) = B(N^2)$$

If B is the constant of proportionality, $P \propto N^2$. Identifying the dimensions of the constant B from $P = B(N^2)$. Thus

$$(ML^{-1} \cdot T^{-2}) = B(T^{-2})$$

from which the dimensions of the numerical constant $B = (ML^{-1})$.
Let

$$P_w = \phi(N) \tag{A2.4}$$

then $P_w = B(N^a)$. From Table A1.1, the dimensions of the terms are: $(ML^2 \cdot T^{-3}) = B(T^{-1})$. Then

$$(ML^2 \cdot T^{-3}) = T^{-a}$$

For M, $1 = 0$;
for L, $2 = 0$;
for T, $-3 = -a$, from which $a = 3$.

Substituting into equation A2.4:

$$P_w = \phi(N^3) = B(N^3)$$

If B is the constant of proportionality, $P_w \propto N^3$. Identifying the dimensions of the constant B from $P_w = B(N^3)$. Thus

$$ML^2 \cdot T^{-3} = B(T^{-3})$$

from which the dimensions of the numerical constant $B = (ML^2)$.

SUMMARISING EXAMPLE A2.1
Table A2.1 confirms the relationship between the variables in Example A2.1 and identifies the dimensions of B, the numerical constants in the solutions. In order to remove the dimensions of B it is necessary to consider what other variables may contribute to each of the three equations for Q, P and P_w. There are a number of possibilities when considering prime movers such as centrifugal pumps and fans. They include:

impeller diameter d;
impeller speed N;
fluid density ρ;
absolute viscosity of the fluid μ;
kinematic viscosity v.

By analysing the dimensions of each of these variables from Table A1.1 (the dimension for impeller diameter d being L) it is possible to deduce the

Table A2.1 Summary of Example A2.1; the laws for pumps and fans

Relationship between variables	Dimensions of numerical constant B	Equivalent terms	Relationship when constant B is dimensionless
$Q \propto N$	L^3	d^3	$Q \propto N \cdot d^3$
$P \propto N^2$	ML^{-1}	$\rho \cdot d^2$	$P \propto N^2 \cdot d^2 \cdot \rho$
$P_w \propto N^3$	ML^2	$\rho \cdot d^5$	$P_w \propto N^3 \cdot d^5 \cdot \rho$

term or terms whose dimensions will equate with the numerical constant B. These are shown in the third column of Table A2.1. You should confirm that the equivalent terms under column three in Table A2.1 have dimensions which will cancel the dimensions of the numerical constant in column two of the table. Thus the numerical constant B is now reduced to a dimensionless number in each case and therefore:

$$Q = (B)Nd^3, \quad P = (B)N^2 \cdot d^2 \cdot \rho \quad \text{and} \quad P_w = (B)N^3 \cdot d^5 \cdot \rho$$

The following example shows a more detailed analysis of the relationship of P_w to N.

Example A2.2 Show the dimensionless groups in an equation for the power of a fan

(a) Show by dimensional analysis that the power P_w required by a fan of diameter d rotating at speed N and delivering a volume per unit time Q of a fluid density ρ and viscosity μ is given by: $P_w = (\rho N^3 d^5), \phi[(Nd^3/Q), (\rho Nd^2/\mu)]$.
(b) Identify the dimensionless groups in the formula.

Solution (a)
Following the procedure in Case study A2.1

$$P_w = B[(d^a)(N^b)(Q^c)(\rho^d)(\mu^e)] \tag{A2.5}$$

Assuming an exponential relationship and substituting dimensions for the variables using Table A1.1 for reference:

$$ML^2T^{-3} = B[(L)^a(T^{-1})^b(L^3T^{-1})^c(ML^{-3})^d(ML^{-1} \cdot T^{-1})^e]$$

For M, $1 = d + e$
for L, $2 = a + 3c - 3d - e$
for T, $-3 = -b - c - e$

thus

$$d = 1 - e \quad \text{and} \quad a = 2 - 3c + 3(1 - e) + e,$$

from which

$$a = 5 - 3c - 2e \quad \text{also} \quad b = 3 - c - e.$$

Substituting the indicial equations for a, b, and d into equation A2.5

$$P_w = B[(d^{(5-3c-2e)})(N^{(3-c-e)})(Q^c)(\rho^{(1-e)})(\mu^e)]$$

The indices identify three groups: numerical, $-c$ and $-e$. Putting the terms into these groupings, we have:

$$P_w = B[(\rho N^3 d^5), (Nd^3/Q)^{-c}, (\rho Nd^2/\mu)^{-e}]$$

Note that in the last group which has the index $-e$, μ goes in the denominator since its index is $+e$. This is also the case for Q in the second group. Note here that the constant B and the indices $-c$ and $-e$ can be replaced with the 'function of' term ϕ. Thus

$$P_w = (\rho N^3 d^5), \phi[(Nd^3/Q), (\rho Nd^2/\mu)]$$

We have therefore shown that the relationship of the selected variables for determining fan power is correct.

Solution (b)

Applying Buckingham's Pi theorem, there are six variables and three dimensions. Thus there are $(n - m) = (6 - 3) = 3$ dimensionless groups. By analysing the dimensions of the second and third group on the RHS of the formula these are found to be dimensionless. You should now confirm that this is so. If the first group on the right hand side $(\rho N^3 d^5)$ is taken over to the left, the third dimensionless group of variables is obtained and will therefore be $(P_w/\rho N^3 d^5)$ and therefore

$$(P_w/\rho N^3 d^5) = \phi[(Nd^3/Q), (\rho Nd^2/\mu)]$$

You should now confirm that this also is the case. Furthermore, the constant B in equation 2.5 must be dimensionless since all three groups here are dimensionless.

Example A2.3 Find the scale and corresponding speed of a model for a given fan performance

The performance of a fan is to be estimated using a scale model. The prototype fan duty is 4 m^3/s at 8 revs/s and the power absorbed is 1450 W whereas the model absorbs 800 W for a flow of 1.3 m^3/s. Determine:

(a) The scale of the model.
(b) The corresponding speed of the model.

Assume the density and viscosity of the air is constant.

Appendix 2   249

Solution (a)
Adopting the formula for power P_w required by the fan in Example A2.2 and using the dimensionless groups identified in part (b) of the solution to Example A2.2.

$$(d^3 N/Q)_m = (d^3 N/Q)_{fs}$$

thus

$$N_m = (Q_m/Q_{fs})N_{fs}(d_{fs}^3/d_m^3)$$

Substituting:

$$N_m = (1.3/4) \times 8(d_{fs}^3/d_m^3)$$

from which

$$N_m = 2.6(d_{fs}^3/d_m^3)$$

Also

$$(P_w/d^5 N^3)_m = (P_w/d^5 N^3)_{fs}$$

then

$$(P_m/P_{fs})N_{fs}^3 = N_m^3(d_m^5/d_{fs}^5)$$

Substituting:

$$(800/1450) \times 8^3 = N_m^3(d_m^5/d_{fs}^5)$$

Substituting for N_m:

$$282.48 = (2.6 d_{fs}^3/d_m^3)^3 (d_m^5/d_{fs}^5)$$

from which

$$282.48/17.576 = (d_{fs}^9/d_m^9)(d_m^5/d_{fs}^5)$$

and therefore

$$16.072 = (d_{fs}^4/d_m^4)$$

thus

$$2 = (d_{fs}/d_m)$$

and the prototype is therefore twice the size of the model.

Solution (b)
From $N_m = 2.6(d_{fs}^3/d_m^3)$

$$N = 2.6(2/1)^3 = 20.8 \text{ revs/s}$$

Example A2.4 Verify the dimensionless groups for the wall shear stress for the mass transfer of fluid in a pipe

For a fluid flowing in a pipe, the wall shear stress τ is considered to be dependent upon the following variables: pipe diameter d, fluid density ρ, fluid viscosity μ, the mean fluid velocity u and the mean height of the roughness projections on the pipe wall k_s

(a) Show by dimensional analysis that $(\tau/\rho u^2) = \phi[(\rho u d/\mu), (k_s/d)]$.
(b) Identify the dimensionless groups in the formula.

Solution (a)
Assuming an exponential relationship where B is the constant of proportionality and the indices a, b, c, d and e are numerical constants:

$$\tau = B[(d^a)(\rho^b)(\mu^c)(u^d)(k_s^e)] \tag{A2.6}$$

Shear stress $\tau = $ (force/area) and therefore has the same dimensions as pressure. Refer to Table A1.1 for this and the dimensions of the other terms. Substituting the dimensions of the variables and the appropriate indices into equation A2.6:

$$(ML^{-1} \cdot T^{-2}) = B[(L^a)(M^b L^{-3b})(M^c L^{-c} T^{-c})(L^d T^{-d})(L^e)]$$

The indicial equations are:

For M, $1 = b + c$
for L, $-1 = a - 3b - c + e + d$
for T, $-2 = -c - d$

From which

$$b = 1 - c \quad \text{and} \quad d = 2 - c.$$

Now

$$a = 3(1-c) + c - e - d - 1$$
$$= 3 - 3c + c - e - d - 1$$
$$= 3 - 2c - e - (2-c) - 1$$
$$= 3 - 2c - e - 2 + c - 1$$

and finally,

$$a = -c - e.$$

Substituting the indicial equations for a, b and d into equation A2.6

$$\tau = B[(d^{(-c-e)}), (\rho^{(1-c)}), (\mu^c), (u^{(2-c)}), (k_s^e)]$$

There are three indices here: numerical, c and e and therefore three groups of variables, thus

$$\tau = B[(u^2 \cdot \rho), (\rho u d/\mu)^{-c}, (k_s/d)^e]$$

and

$$\tau = \phi[(u^2 \cdot \rho), (\rho u d/\mu), (k_s/d)]$$

From this expression you will recognise the middle group as the Reynolds number and the last group as relative roughness (see Section 6.5 of Chapter 6 for the introduction of this last term), both of which are dimensionless.

The formula for shear stress can therefore be written as:

$$\tau = (u^2 \rho), \phi[(\rho u d/\mu), (k_s/d)]$$

Solution (b)
Since the second and third groups are dimensionless, by taking the first group over to the left-hand side, the remaining dimensionless group will be: $(\tau/u^2 \rho)$. Buckingham's Pi theorem of $(n-m) = (6-3) = 3$, therefore, agrees with this solution. You should now confirm that the three groups of terms are dimensionless.

Example A2.5 Find the form of the equation for forced convection from a fluid transported in a straight pipe

The heat transfer by forced convection from a fluid transported in a long straight tube is governed by the variables h, d, μ, ρ, k, C, θ and u such that the heat transfer coefficient for forced convection $h = \phi$ (d, μ, ρ, k, C, θ, u).

252 Appendix 2

Using dimensional analysis, determine the form of the equation and the dimensionless grouping.

Solution
If an exponential relationship is assumed in which B is the constant of proportionality and the indices a, b, c, e, f and g are numerical constants,

$$h = B[(d^a)(\mu^b)(\rho^c)(k^e)(C^f)(u^g)] \tag{A2.7}$$

Introducing dimensions to each of the terms using Table A1.1 and including the indices as appropriate:

$$(MT^{-3} \cdot \theta^{-1}) = B[(L^a), (M^b \cdot L^{-b} \cdot T^{-b}), (M^c \cdot L^{-3c}),$$
$$(M^e \cdot L^e \cdot T^{-3e} \cdot \theta^{-e}), (L^{2f} \cdot T^{-2f}\theta^{-f}), (L^g \cdot T^{-g})$$

Collecting the indices:
For M,

$$1 = b + c + e \tag{A2.7v}$$

For L,

$$0 = a - b - 3c + e + 2f + g \tag{A2.7w}$$

For T,

$$-3 = -b - 3e - 2f - g \tag{A2.7y}$$

For θ,

$$-1 = -e - f \tag{A2.7z}$$

There are six unknown indices and four indicial equations. Evaluating in terms of the unknown indices c and f. From equation A2.7z,

$$e = 1 - f \tag{A2.7r}$$

Substitute equation A2.7r into equation A2.7v,

$$1 = b + c + 1 - f$$

from which

$$b = f - c \tag{A2.7s}$$

Substitute equations A2.7r and A2.7s into A2.7y,

$$-3 = -(f-c) - 3(1-f) - 2f - g$$
$$-3 = -f + c - 3 + 3f - 2f - g$$

from which

$$g = c \qquad (A2.7t)$$

Substitute equations A2.7r, A2.7s and A2.7t into equation A2.7w

$$0 = a - (f - c) - 3c + (1 - f) + 2f + c$$
$$0 = a - f + c - 3c + 1 - f + 2f + c$$

from which

$$a = c - 1$$

Substituting the indicial equations for a, b, g and e into equation A2.7

$$h = B[(d^{(c-1)}), (\mu^{(f-c)}), (\rho^c), (k^{(1-f)}), (C^f), (u^c)]$$

The variables are now related to the unknown indices c and f and index 1.0. There are therefore three groups of variables. Thus

$$h = B[(k/d), (d\rho u/\mu)^c, (\mu C/k)^f]$$

Adopting Buckingham's Pi theorem there are $(n - m) = (7 - 4) = 3$ dimensionless groups here, thus by rearranging the equation they are:

$$(hd/k) = B[(d\rho u/\mu)^c, (\mu C/k)^f]$$

The group (hd/k) is known as the Nusselt number Nu; the group $(d\rho u/\mu)$ is the Reynolds number Re and the group $(\mu C/k)$ is known as the Prandtl number Pr.

The value of the numerical (dimensionless) constants B, c and f are found empirically (by experiment). It has been established that the values of B, c and f are constant for a very wide range of Re and Pr numbers and $B = 0.023$, $c = 0.8$ and $f = 0.33$. The heat transfer correlation for flow of a fluid through a long tube is therefore:

$$Nu = 0.023(Re)^{0.8}(Pr)^{0.33}$$

This formula is introduced in Chapter 3 as equation 3.10. You should now confirm that the Nusselt, Reynolds and Prandtl numbers are dimensionless.

Example A2.6 Find the form of the equation for free convection in turbulent flow over vertical plates

Heat transfer by free convection in turbulent flow over vertical plates is governed by the variables h, dt, β, g, d, ρ, μ, k and C such that the heat transfer coefficient for free convection $h = \phi(dt, \beta, g, d, \rho, \mu, k, C)$. Using dimensional analysis, determine the form of the equation and the dimensionless groups.

Solution

If an exponential relationship is assumed in which B is the constant of proportionality and the indices a, b, e, f, j and n are numerical constants,

$$h = B[((dt \cdot \beta g)^a), (d^b), (\rho^e), (\mu^f), (k^j), (C^n)] \tag{A2.8}$$

Introducing dimensions to each of the terms using Table A1.1 with the exception of the terms dt, β and g which are combined as shown in equation A2.8 to ensure that the unknown indices are limited to six and including the indices as appropriate:

$$MT^{-3} \cdot \theta^{-1} = B[(L^a \cdot T^{-2a}), (L^b), (M^e \cdot L^{-3e}), (M^f \cdot L^{-f} \cdot T^{-f}),$$
$$(M^j \cdot L^j \cdot T^{-3j} \cdot \theta^{-j}), (L^{2n} \cdot T^{-2n} \cdot \theta^{-n})$$

Collecting the indices:
For M,

$$1 = e + f + j \tag{A2.8p}$$

For L,

$$0 = a + b - 3e - f + j + 2n \tag{A2.8q}$$

For T,

$$-3 = -2a - f - 3j - 2n \tag{A2.8r}$$

For θ,

$$-1 = -j - n \tag{A2.8s}$$

There are six unknown indices and four indicial equations. Evaluating in terms of the unknown indices a and n.
From equation (A2.8p)

$$f = 1 - e - j \tag{A2.8t}$$

from equation A2.8s,

$$j = 1 - n \tag{A2.8u}$$

Substitute equations A2.8t and A2.8u into equation A2.8r

$$-3 = -2a - (n - e) - 3(1 - n) - 2n$$

thus

$$-3 = -2a - n + e - 3 + 3n - 2n$$

and

$$e = 2a \tag{A2.8v}$$

Substitute equations A2.8s, A2.8t and A2.8u into equation A2.8q

$$0 = a + b - 3e - (1 - e - j) + (1 - n) + 2n$$

thus

$$0 = a + b - 3e - 1 + e + j + 1 - n + 2n$$

Substitute equation A2.8u

$$0 = a + b - 2e + (1 - n) + n$$

Substitute equation A2.8v

$$0 = a + b - 4a + 1$$

from which

$$b = 3a - 1$$

Substitute equations A2.8u and A2.8v into equation A2.8t

$$f = 1 - 2a - (1 - n)$$

from which

$$f = 1 - 2a - 1 + n$$

and therefore

$$f = n - 2a$$

Substituting the indicial equations for b, e, f and j into equation A2.8

$$h = B[((dt \cdot \beta g)^a), (d^{(3a-1)}), (\rho^{2a}), (\mu^{(n-2a)}), (k^{(1-n)}), (C^n)]$$

The variables are now related to the unknown indices a and n and numerical index 1.0. There are therefore three groups of variables. Thus,

$$h = B[(k/d), (dt\beta g d^3 \cdot \rho^2/\mu^2)^a, (\mu C/k)^n]$$

Adopting Buckingham's Pi theorem there are $(n - m) = (7 - 4) = 3$ dimensionless groups here, thus:

$$(hd/k) = B[(dt \cdot \beta g d^3 \cdot \rho^2/\mu^2)^a, (\mu C/k)^n]$$

where (hd/k) is the Nusselt number Nu, $(dt \cdot \beta g d^3 \cdot \rho^2/\mu^2)$ is the Grashof number Gr and $(\mu C/k)$ is the Prandtl number Pr.

The value of the numerical (dimensionless) constants B, a and n are found empirically. From Chapter 3, equations 3.6 (and 3.14) for air uses the constants $B = 0.13$, 0.1, $a = 0.33$ and $n = 0.33$ when $Gr > 10^9$. Thus

$$Nu = 0.13[(Pr)(Gr)]^{0.33}$$

You will notice that the number of variables n in Buckingham's Pi theorem assumes that the variables dt, β and g are kept together as one, as they are so designated at the commencement of this solution. You should now confirm that the Nusselt, Prandtl and Grashof numbers are dimensionless.

SUMMARY FOR EXAMPLES A2.5 AND A2.6

In both of these solutions you can see that the formulae for forced convection inside long tubes and free turbulent convection over vertical plates (free turbulent convection over vertical cylinders) only require experimental work to evaluate the numerical and indicial constants. Dimensional analysis therefore can provide a fast track methodology for deriving empirical formulae in which the variables have been identified. These rational formulae would have been generated by the use of dimensional analysis as described above.

The following example employing dimensional analysis relates to heat flow into a wall.

Example A2.7 Find the form of an equation for the rise in surface temperature of a wall, the dimensionless groups in the equation and the heat up time

A cavity wall constructed from two leaves of brick is subjected to a heat flux of I W/m^2 for time T seconds. The temperature rise $d\theta$ at the inside surface depends upon the physical properties of the lining which are its thermal conductivity k, density ρ and specific heat capacity C. The formula is likely to be in the form $d\theta = \phi[(I, T, k, \rho, C]$

(a) By employing dimensional analysis find the form of the equation which will determine the rise in surface temperature of the wall after time T seconds.
(b) By adopting Buckingham's Pi theorem establish the dimensionless groups in the formula.
(c) Given that the numerical constant of proportionality B is $(2/\pi^{0.5})$, determine how long it will take to raise the inner surface temperature of the wall by 4 K. The width of the inner brick leaf is 100 mm and the temperature drop between its inner and outer surface is 3 K.

Take the thermal conductivity of the inner brick leaf of the wall as 0.62 W/mK, density 1700 kg/m³ and specific heat capacity 800 J/kgK. See Table A2.1.

Solution (a)
If an exponential relationship is assumed where B is the constant of proportionality and the indices a, b, c, d and e are numerical constants,

$$d\theta = B[(I^a), (T^b), (k^c), (\rho^d), (C^e)] \tag{A2.9}$$

The dimensions of the variables in the proposed formula can be obtained from Table A1.1. In this solution, however, the additional dimension H for heat energy in Joules will be used. Thus heat flux in J/s m² will have the dimensions $(HT^{-1} \cdot L^{-2})$. Likewise thermal conductivity in J/s mK will have the dimensions $(HT^{-1} \cdot L^{-1} \cdot \theta^{-1})$ and specific heat capacity in J/kgK will have the dimensions $(HM^{-1} \cdot T^{-1})$. There will now be five dimensions used in this solution, namely H, M, L, T, θ. Introducing dimensions to each of the terms in equation A2.9 and including the indices where appropriate:

$$d\theta = B[(H^a \cdot T^{-a} \cdot L^{-2a}), (T^b), (H^c \cdot T^{-c} \cdot L^{-c} \cdot \theta^{-c}),$$
$$(M^d \cdot L^{-3d}), (H^e \cdot M^{-e} \cdot \theta^{-e})]$$

Collecting the indices:
For H,

$$0 = a + c + e \tag{A2.9m}$$

For M,

$$0 = d - e \tag{A2.9n}$$

For L,

$$0 = -2a - c - 3d \tag{A2.9o}$$

258 Appendix 2

For T,

$$0 = -a + b - c \tag{A2.9p}$$

For θ,

$$1 = -c - e \quad \text{also } e = -1 - c \tag{A2.9q}$$

From equation A2.9n,

$$d = e \tag{A2.9r}$$

From equation A2.9p,

$$a = b - c \tag{A2.9s}$$

Substitute equations A2.9q and A2.9s into equation A2.9m:

$$0 = b - c + c - 1 - c$$

From which

$$1 = b - c \tag{A2.9t}$$

Substitute equation A2.9t into equation A2.9p,

$$0 = -a + 1$$

From which

$$a = 1 \tag{A2.9u}$$

Add equations A2.9m and A2.9o

$$0 = -a - 3d + e$$

Substitute equation A2.9r:

$$0 = -a - 3e + e$$
$$0 = -a - 2e$$

Substitute equation A2.9u

$$0 = -1 - 2e$$

From which

$$e = -0.5 = d \tag{A2.9v}$$

Appendix 2   259

Substitute equation A2.9v in equation A2.9q

$$1 = -c - (-0.5)$$

From which

$$c = -0.5 \qquad (A2.9w)$$

Substitute equations A2.9u and A2.9w into equation A2.9p

$$0 = -1 + b - (-0.5)$$

from which

$$b = 0.5 \qquad (A2.9x)$$

The indices a, b, c, d and e each have numerical values as identified in equations A2.9u, v, w, x. Substituting these indicial values into equation A2.9:

$$d\theta = B(I \cdot T^{0.5} \cdot k^{-0.5} \cdot \rho^{-0.5} \cdot C^{-0.5})$$

thus

$$d\theta = B[(I)(T/k\rho C)^{0.5}]$$

In this solution the only numerical value which must be found empirically is the constant B since the numerical values of the indices a, b, c, d and e have been evaluated during the process of analysis.

Solution (b)
From Buckingham's Pi theorem $(n - m) = (6 - 5) = 1$ dimensionless group. Rearranging the formula derived in part (a):

$$1 = B[(I/d\theta)(T/k\rho C)^{0.5}] \qquad (A2.10)$$

The dimensions of the terms are now analysed.

The dimensions of the terms $(I/d\theta)$ are: $(HT^{-1} \cdot L^{-2} \cdot \theta^{-1})$.
The dimensions of the terms $(T/k\rho C)^{0.5}$ are:
$(TH^{-1} \cdot TL\theta M^{-1} \cdot L^3 \cdot H^{-1} \cdot M\theta)^{0.5}$.
This reduces to: $(T^2 \cdot H^{-2} \cdot L^4 \cdot \theta^2)^{0.5}$. Accounting for the index of 0.5, the dimensions of the terms $(T/k\rho C)^{0.5}$ are : $(TH^{-1} \cdot L^2 \cdot \theta)$.

Now combining the dimensions of the terms: $(I/d\theta)(T/k\rho C)^{0.5}$ we have: $(HT^{-1} \cdot L^{-2} \cdot \theta^{-1} \cdot TH^{-1} \cdot L^2 \cdot \theta)$ from which this group is dimensionless.

260 Appendix 2

SUMMARY OF PARTS (a) AND (b) OF EXAMPLE A2.7
You can see by a process of cancellation, the combined group of terms in equation A2.10 is dimensionless and Buckingham's Pi theorem confirms this. The numerical constant B is also therefore dimensionless. The index of 0.5 in the formula is accounted for in the determination of the dimensionless group since it was evaluated during the process of deriving the equation. In the case of Examples A2.5 and A2.6 the indices have to be found empirically and therefore do not form part of the analysis of the dimensionless groups.

Solution (c)

Adopting the formula obtained in part (a) and using the given constant of proportionality B:

$$d\theta = (2/\pi^{0.5})(I)(T/k\rho C)^{0.5}$$

Rearranging the equation in terms of time T in seconds by first of all squaring both sides:

$$(d\theta)^2 = (4/\pi)(I)^2(T/k\rho C)$$

Thus

$$T = [(d\theta)^2 \cdot \pi k\rho C]/(4 \cdot I^2)$$

Heat flux I, from Fourier's equation (see Section 2.2, Chapter 2) will be:

$$I = k \cdot dt/L = (0.62 \times 3)/0.1 = 18.6 \text{ W/m}^2$$

Substituting for a 4 K rise in wall surface temperature:

$$T = (4^2 \cdot \pi \times 0.62 \times 1700 \times 800)/(4 \times (18.6)^2)$$

from which

$$T = 30\,628 \text{ s}$$

and the time taken to raise the inner surface temperature of the inner brick leaf of the wall by 4 K will be 8 h 30 min.

SUMMARY FOR PART (c) OF EXAMPLE A2.7
Clearly, this is a long heat up period for the plain inner brick wall. If, however, the wall is lined with a proprietary lining 15 mm thick, having a thermal conductivity of 0.06 W/mK, a density of 300 kg/m³ and a specific heat capacity of 1000 J/kgK and a 3 K temperature difference between its faces, heat flux I comes to 12 W/m² and time T evaluates to 1571 s or 26.18 min. You should now confirm this solution using the formulae above.

Thus, thermal insulation on the inside surface of the external envelope of a building provides a short heat up period. Example 2.1 analyses the heat flux I through an external wall under steady conditions. Table 2.1 gives properties of building and thermal insulation materials.

Example A2.8 Verify the form of Pole's formula for natural gas flow in a pipe

Pole's formula for natural gas flow from *CIBSE Guide G* is given as:

$$Q = 0.001978 d^2 (Hd/SL)^{0.5} \text{ L/s}$$

where pipe diameter d is in mm, S is specific gravity which is a ratio of gas to air density each in kg/m^3, L is the pipe length in m, H is the pressure drop in mbar. Show by dimensional analysis that Q the gas flow rate is controlled by the variables d, H, L and S.

Solution
Let

$$Q = B(d^a, H^b, L^c, S^d) \quad (A2.11)$$

where B is the constant of proportionality and a, b, c and d may need to be evaluated empirically. Referring to Table A1.1 for the dimensions of the variables and ignoring for the moment the dimensions for Q, d and H in the question:

$$L^3 T^{-1} = B(L^a, (M^b L^{-b} T^{-2b}), L^c, (M^d L^{-3d}))$$

Forming equations from the indices:

$$M: \quad 0 = b + d$$

So,

$$d = -b$$
$$L: \quad 3 = a - b + c - 3d$$
$$T: \quad -1 = -2b$$

So,

$$b = \tfrac{1}{2} \quad \text{and} \quad d = -\tfrac{1}{2}$$

Substituting into the indicial equation for dimension L:

$3 = a - \frac{1}{2} + c + 1\frac{1}{2}$

$3 = a + c + 1$

$2 = a + c$

$a = 2 - c$

Substituting the indicial equations into equation A2.11 above

$$Q = B(d^{(2-c)}, H^{0.5}, L^c, S^{-0.5})$$

Putting the variables into their groups:

$$Q = B[(d^2), (d/L)^{-c}, ((H/S)^{0.5})]$$

The index $(-c)$ is evaluated to 0.5 by experiment, so:

$$Q = B[d^2(dH/SL)^{0.5}]$$

This confirms that the flow rate Q is controlled by the variables d, H, L and S. The constant B accounts in part for the units of Q in L/s instead of m³/s, pressure H in mbar in lieu of Pa and pipe diameter d in mm instead of m.

A2.5 Appendix closure

Successful completion of this chapter provides the reader with underpinning knowledge in respect of some of the formulae employed in fundamental calculations relating to the design processes in heat and mass transfer in the subjects of heating, ventilating and air conditioning. Dimensional analysis is not a wonder tool. You will have perhaps noticed in the examples selected that a knowledge of the processes to be analysed is essential when it comes to finding the form of an equation relating to a number of variables and this invariably requires a process of iteration or trial and error when starting out. Example A2.5, forced convection in a pipe, and Example A2.6, free convection over vertical plates, demonstrate how these empirical formulae were developed for convective heat flow.

Appendix 3

Renewable energy systems

A3.1 Introduction

Renewable energy now forms part of the design brief for new building projects and refurbishment projects for existing buildings. It is therefore important that basic underpinning knowledge of systems that produce energy, largely free from green house gases and used as energy sources, forms an integral part of our understanding as specialists in the built environment.

For this reason a selection of low to zero carbon technologies (LZC) will form the focus of this appendix. They include

- wind turbines,
- hydro power,
- marine turbines,
- solar thermal heating,
- photovoltaics,
- biomass,
- combined heat and power (CHP),
- fuel cell CHP.

At present CHP is not strictly a renewable energy but it has its place in this list for reasons that will become apparent.

Consideration is being given to moving away from relying exclusively on centralised generation of electricity to *distributed energy systems* in towns and communities – even for single households. There are reasons for this change in direction not least since it reduces transmission losses when the supply is generated and used locally. Clearly the National Grid is needed to pick up power that is surplus or deliver electricity to customers when required. The use of hydro power, marine turbines and wind turbines, for example, can contribute to the needs of the wider community via the National Grid.

There is much discussion on the benefits of onland, onshore and off-shore wind turbines. However, large numbers are now in use and providing a small but significant power supply to commercial and industrial sites and communities local to the wind farms.

A3.2 Wind turbines

The use of the wind to provide the motive force for machinery has been in use for centuries starting with the wind mill. The use of the wind to generate electricity is a more recent venture.

The most common type of wind turbine has a horizontal rotor although vertical rotor turbines are in very limited use as a consequence of relatively low maximum power output and difficulties in locating suitable sites. The horizontal rotor machines are in use in the United Kingdom and located inland, onshore and offshore. They generate DC current which is converted to AC current on the farm or site for local use and connection to the grid.

There is public concern over the location of suitable sites, particularly for wind farms. This is due to a number of reasons, one of which is the noise generated by the turbine. Turbines are now available with no gearing mechanism, the main cause of noise generation, as a result of flexible blade technology that allows the turbine to continue operating in high wind speeds. This increases the overall efficiency since the turbine blades are not feathered in high winds.

From Section 7.2, kinetic energy or velocity pressure generated by the air which is the prime mover of the turbine blades is $P_u = 1/2 u^2 \rho$ Pa. If the area swept by the blades is represented as a_s then the volume flow rate (Q) of air over the blades will be

$$Q = a_s \cdot u \text{ m}^3/\text{s}$$

Mean air velocity (u) at rotor hub height will vary from a maximum beyond which the blades are turned away from the wind to a minimum which will be zero. The theoretical power (P_w) generated by the prevailing wind velocity is obtained from $P_w = Q \cdot P_u$ Watts. This may be analysed with reference to Table A1.1, where the dimensions for power (P_w) are ML^2T^{-3}. The dimensions of volume flow rate (Q) are L^3T^{-1} and the dimensions of velocity pressure (P_u) are $ML^{-1}T^{-2}$. The product of the dimensions of these two terms is $(L^3T^{-1})(ML^{-1}T^{-2})$. This reduces to ML^2T^{-3} which agrees with the dimensions for power (P_w) above and in Table A1.1.

From above,

$$P_w = Q \times P_u \text{ Watts}$$

Thus theoretical power generated by the turbine

$$P_w = (a_s \cdot u)\left(\tfrac{1}{2} \cdot u^2 \cdot \rho\right)$$

And therefore theoretical power

$$P_w = \tfrac{1}{2} \cdot a_s \cdot u^3 \cdot \rho \text{ Watts} \tag{A3.1}$$

Appendix 3   265

Annual energy (H) in kWh can be evaluated using P_w in Watts. Thus

theoretical $H = P_w \times (8736/1000)$ kW

where there are 8736 h in 1 year and there are 1000 W in 1 kW. Substituting for P_w we have

$$H = \left(\tfrac{1}{2} \cdot a_s \cdot u^3 \rho\right)(8736/1000)$$

Taking air density $= 1.2$ kg/m³, then theoretical energy

$$H = (5.24 a_s \cdot u^3) \text{ kWh/annum} \tag{A3.2}$$

There are three efficiencies that apply to wind turbines:

$\eta_g =$ generator efficiency which is around 0.95.

$\eta_r =$ rotor efficiency which is in the region of 0.45.

$\eta_u =$ efficiency of utilisation that is in the region of 0.85.

Efficiencies (η_g) and (η_r) apply to the calculation of potential turbine power (P_w) and efficiencies (η_g), (η_r) and (η_u) apply to the determination of potential annual energy output (H).

There now follows an example related to a wind farm serving a housing estate.

Example A3.1 Find the annual energy generated from a wind farm, the surplus energy available to the National Grid, and the carbon dioxide emissions saved

A wind farm has a total of 10 turbines each rated at 300 kW. The cut in speed is 3 m/s and the average wind speed may be taken as 7.5 m/s at rotor hub height. Rotor diameter is 37 m and hub height is 35 m. The wind farm serves 3000 houses.

(a) Find the wind speed on which the output rating is based. Adopt a generator efficiency of 95% and a rotor efficiency of 45%.
(b) Find the annual energy generated by the wind farm in MWh based on the mean wind speed. Adopt a utilisation efficiency of 85%.
(c) If the average annual electricity consumption of each house is 2000 kWh calculate the annual surplus energy in MWh available from the wind farm to the National Grid.
(d) Estimate the annual carbon dioxide emissions in tonnes that would pass into the atmosphere if a power station were to generate the corresponding annual output of the wind farm. Use the conversion factor of 0.43 kgCO$_2$/kWh.

Solution (a)
From equation A3.1, practical power

$$P_\text{w} = \tfrac{1}{2} a_\text{s} \cdot u^3 \cdot \rho \cdot \eta_\text{g} \cdot \eta_\text{r} \ \text{W}$$

Rearranging in terms of velocity

$$u = \sqrt[3]{(2P_\text{w})/(a_\text{s} \cdot \rho \eta_\text{g} \cdot \eta_\text{r})}$$

Substituting

$$u = \sqrt[3]{[(2 \times 300\,000)/(1075 \times 1.2 \times 0.95 \times 0.45)]}$$

From which

$$u = 10.3 \ \text{m/s}$$

Solution (b)
From equation A3.2 practical energy

$$H = 5.24 a_\text{s} \cdot u^3 \eta_\text{g} \cdot \eta_\text{r} \cdot \eta_\text{u} \ \text{kWh/annum}$$

Substituting, practical energy

$$H = 5.24 \times 1075 \times (7.5)^3 \times 0.95 \times 0.45 \times 0.85$$

From which

$$H = 863\,532 \ \text{kWh/annum}$$

For 10 turbines

$$H = 8\,635\,320 \ \text{kWh/annum} = 8635 \ \text{MWh/annum}$$

Solution (c)
Available energy to each house

$$H = (8\,635\,320)/(3000) = 2878 \ \text{kWh}$$

Surplus energy

$$H = 3000(2878 - 2000) = 2\,634\,000 \ \text{kWh} = 2634 \ \text{MWh}$$

This energy would be fed into the National Grid annually.

Solution (d)
Annual carbon dioxide emissions from a traditional power station

$$\text{annual CO}_2 = 8\,635\,320 \times 0.43 = 3\,713\,187 \text{ kg} = 3713 \text{ tonnes}$$

SUMMARY FOR EXAMPLE A3.1
The solutions above are reliant on the wind speed (u) to the power of 3, so small variations in wind velocity over the turbine blades have a highly significant effect on turbine output and generated energy. Note the savings in CO_2 emissions from using a wind farm.

A3.3 Hydro power

Hydraulic power is a green energy source and has been in use for centuries starting with water mills employed to directly drive machinery.

There are three main types of water turbine. For vertical heads between 150 m and 1900 m the Pelton Impulse Turbine is used consisting of a series of radial buckets. Refer to Figure A3.1. It can run in the vertical or horizontal position. The Francis Reaction Turbine is used for vertical heads of 50–500 m and is usually run with the impeller in the vertical position and the generator mounted to the vertical shaft. Refer to Figure A3.2. The Kaplan Reaction Turbine is employed for vertical heads of up to 70 m. It is run in the vertical position and has propeller blades that are adjustable allowing variable outputs and ensuring maximum efficiency at all loads. Refer to Figure A3.3.

Hydro power is a well-tried method for generating electricity. It includes the use of dams and pumped storage systems. The use of dams to provide the static head (h) of water to drive the turbines is probably the most well-known form of hydro power. Power in Watts can be determined from

$$P_w = M \cdot g \cdot h \text{ W} \tag{A3.3}$$

This formula can be verified by using dimensional analysis that is introduced in Appendix 1. Using the data in Table A1.1, the dimensions of each term on the right-hand side are: $(MT^{-1})(L \cdot T^{-2})(L)$. This reduces to (ML^2T^{-3}) which is the dimension for power (P_w) in Watts.

Clearly it can be seen from equation A3.3 that the static height of the dam above the turbine power house and the rate of mass transfer of water in the penstock have a significant effect on the power being generated. A small static head over a weir requires a large mass transfer of water to generate a decent output.

Example A3.2 Determine the flow of water over a weir to generate 25 kW

Find the theoretical volume flow of water over a 2 m weir that passes through a series of turbines which together generate 25 kW.

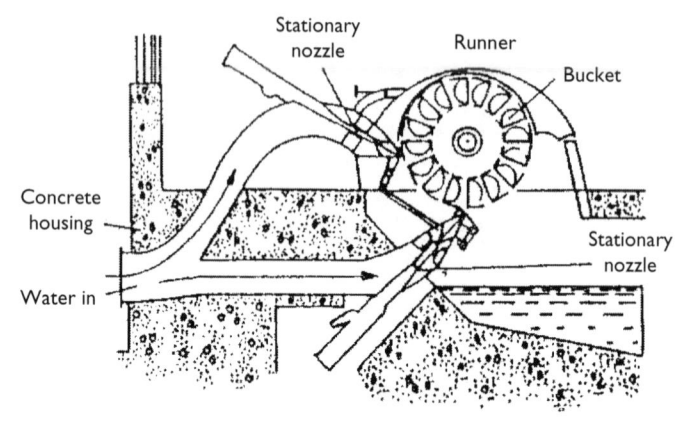

Section through an impulse (Pelton wheel)
type of hydraulic turbine unit
(much simplified)

Figure A3.1 The Pelton Impulse Turbine.

Source: Reproduced with permission of the Heating and Ventilating Contractors' Association.

Cross section through
a Francis type turbine
installation

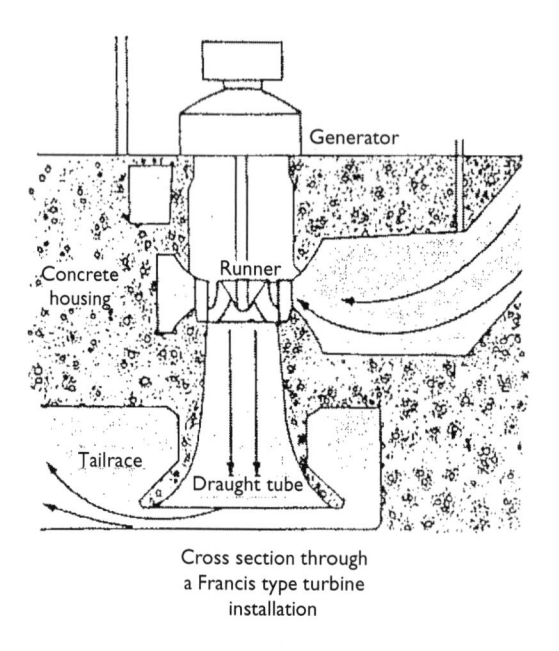

Figure A3.2 The Francis Turbine.

Source: Reproduced with permission of the Heating and Ventilating Contractors' Association.

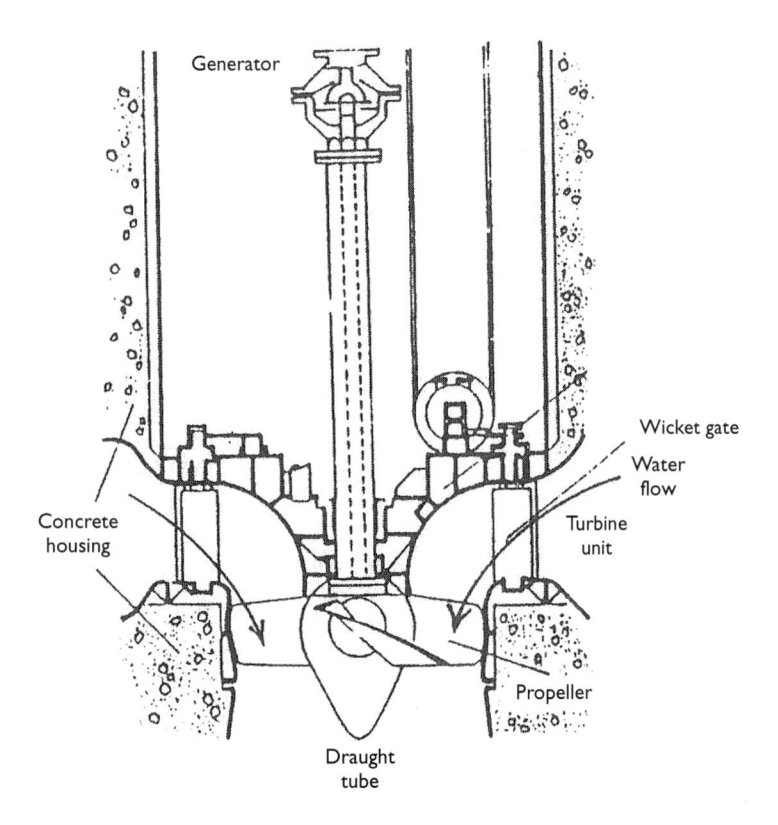

Figure A3.3 Schematic cross section of the Kaplan Turbine.
Source: Reproduced with permission of the Heating and Ventilating Contractors' Association.

Solution
From equation A3.3, $P_w = M \cdot g \cdot h$ Watts. So mass transfer,

$$M = P_w/(gh) = 25\,000/(9.81 \times 2) = 1274 \text{ kg/s}$$

The volume flow of water is, therefore, 1274 L/s or 1.274 m³/s.

Hydro power generated using a high level lake or reservoir may have the generating station located some distance downstream with a penstock connected from the lake and run overland eventually following the river or stream to the generating station. The following example deals with a hydro power station served from a high level reservoir.

Example A3.3 Find the mass transfer of water into a hydro turbine, the turbine power generated and the energy stored in the upper reservoir

Figure A3.4 shows in elevation a hydro power station connected by a penstock 400 m long to a reservoir that is 70 m vertically above it. The hydraulic loss through the Francis turbine located in the power station is 1 bar with negligible losses through the tale race.

There is a bellmouth connection at entry from the lake into the penstock and the coefficient of friction, which is a constant, within the penstock pipe 300 mm diameter, may be taken as 0.004.

(a) Determine the mean velocity of the water flow in the penstock and hence the mass transfer of water handled by the Francis turbine.
(b) Calculate the power output developed by the turbine and generator given efficiencies of 60% and 95%, respectively.
(c) Estimate the energy stored in MWh in the reservoir given its surface area as 4 (km)^2 and usable water depth of 1.0 m.

Solution (a)
In Sections 5.3 and 7.2 the Bernoulli equation is introduced and used. It is relevant here, and from Figure A3.4 the total energy at section 1 is equal to the total energy at section 2, thus:

$$Z_1 + P_1/\rho g + u_1^2/2g = Z_2 + P_2/\rho g + u_2^2/2g + \text{losses metres of}$$
$$\text{fluid flowing}$$

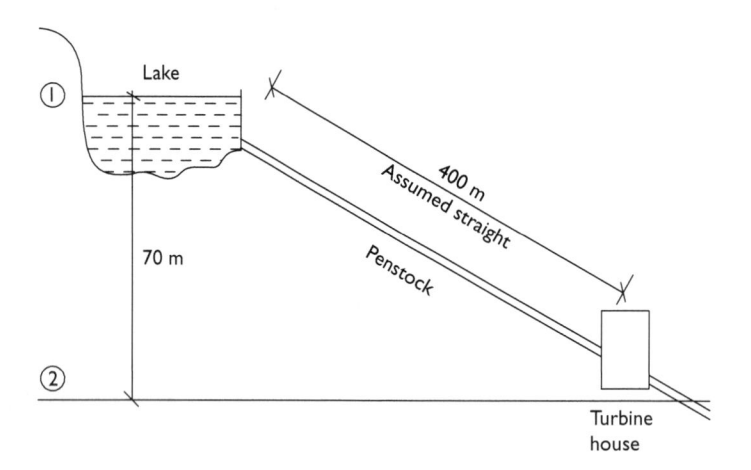

Figure A3.4 Example A3.3 – the hydroelectric system.

Now $P_1 = P_2$ and $u_1 = 0$. Then,

$$Z_1 - Z_2 = u_2^2/2g + \text{losses}$$

The losses include the loss in the penstock (equation 6.5) and the loss in (h) metres through the turbine where $h = P/\rho g = 100\,000/(1000 \times 9.81) = 10.2$ m. The entry connection from the reservoir is bellmouthed and therefore the hydraulic loss is negligible.

Thus for the penstock,

$$h = 4fLu^2/2gd.$$

And substituting:

$$Z_1 - Z_2 = u_2^2/2g + 4fLu_2^2/2gd + 10.2$$
$$70 = u_2^2/2g + (4 \times 0.004 \times 400u_2^2)/(2 \times 9.81 \times 0.3) + 10.2$$
$$= u_2^2/2g + (21.33u_2^2)/2g + 10.2$$

Rearranging and simplifying

$$59.8 = (22.33u_2^2)/2g$$

From which

$$59.8 = 1.383u_2^2 \text{ and } u_2 = 7.25 \text{ m/s}$$

The gravitational volume flow rate of water through the penstock

$$Q = u \cdot a = 7.25 \times \pi(0.3)^2/4 \text{ m}^3/\text{s}$$

So $Q = 0.552$ m³/s $= 552$ L/s

And the gravitational mass transfer of water $M = 552$ kg/s.

Solution (b)
Theoretical power $P_w = M \cdot g \cdot h = 552 \times 9.81 \times 70 = 379\,058$ W
Practical power $= 379\,058 \times 0.6 \times 0.95 = 216\,063$ W $= 216$ kW

Solution (c)
Usable water volume in the reservoir $= 4\,000\,000$ m³
Storage time $= (4\,000\,000)/(0.552 \times 3600) = 2013$ h
Stored energy in the reservoir $H = 216 \times 2013 = 434\,808$ kWh $= 435$ MWh

SUMMARY FOR EXAMPLE A3.3
If the power station was required to supply 216 kW continuously, it would be able to do it for $2013/(24 \times 7) = 12$ weeks without replenishment of water in the reservoir. During the 12 weeks, without replenishment, the vertical height of the water above the power station would be lowered from 70 m to 69 m thus reducing its output.

Hydro power pumped storage

One of the problems of providing electrical energy is that its use is not constant throughout the day and the peaks must be accommodated by switching in extra generating capacity. An alternative is to pump water from a low reservoir to a reservoir located at a much higher level during times of low electrical demand so that the potential energy can be translated into electricity by allowing flow reversal through a turbine when demand builds up.

Example A3.4 Find the gravitational mass transfer of water handled by a hydro turbine and the power developed and calculate the pump duty to return the water to the upper storage reservoir

Figure A3.5 shows in elevation a pumped storage system. The penstock connecting the two lakes having a difference in levels of 150 m is 200 mm in diameter and has a straight length of 180 m and contains one bend having a velocity head loss factor of 1.1. The entry and exit points of the penstock to the lakes are bellmouthed. The coefficient of friction due to water flow in the penstock is 0.005 for both pump and turbine operations.

(a) Determine the mean velocity of gravitational water flow in the penstock and hence the mass transfer of water handled by the turbine.
(b) Calculate power developed by the generator given a turbine efficiency of 70% and a generator efficiency of 95%.

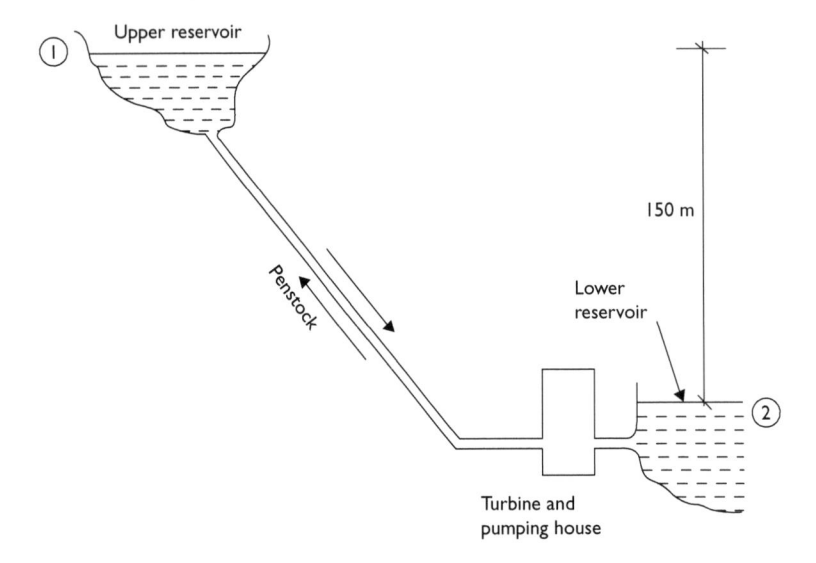

Figure A3.5 Example A3.4 – the pumped storage system.

(c) If the storage pump handles 0.16 m³/s of water from the lower lake find the pressure it must develop for the mass transfer of water to the high level lake and hence state the pump duty.

(d) Calculate the input power required for the storage pump given an overall pump efficiency of 70%.

(e) If the storage pump operates for 7 h during the off peak period, how long can the turbine operate at full load during the peak period in order to maintain the water level in the upper lake.

Solution (a)
In Sections 5.3 and 7.2 the Bernoulli equation is introduced and used. It is relevant here and from Figure A3.5 the total energy at section 1 is equal to the total energy at section 2, thus:

$$Z_1 + P_1/\rho g + u_1^2/2g = Z_2 + P_2/\rho g + u_2^2/2g + \text{losses metres of}$$
$$\text{fluid flowing}$$

Now at the water surfaces $P_1 = P_2$ and $u_1 = u_2$. Then

$$Z_1 - Z_2 = \text{losses in the bend and in the straight pipe}$$

The hydraulic losses in the system $= k(u_2^2/2g) + (4fLu^2/2gd)$
So

$$Z_1 - Z_2 = k(u_2^2/2g) + (4fLu^2/2gd)$$

and substituting

$$150 = 1.1(u_2^2/2g) + (4 \times 0.005 \times 180u^2)/(2g \times 0.2)$$

Rearranging:

$$150(2g) = 1.1u_2^2 + 18u^2$$
$$2943 = 19.1u^2$$

From which mean velocity of the water in the penstock $u = 12.413$ m/s
Gravitational volume flow $Q = u \cdot a = u(\pi d^2/4)$ m³/s
Substituting $Q = 12.413[\pi(0.2)^2/4] = 0.39$ m³/s $= 390$ L/s
The gravitational mass transfer of water in the penstock $M = 390$ kg/s

Solution (b)
Theoretical power

$$Pw = M \cdot g \cdot h = 390 \times 9.81 \times 150 = 573\,885 \text{ W}$$

Practical power developed by the generator

$$P_w = 573\,885 \times 0.7 \times 0.95$$
$$= 381\,633 \text{ W} = 381 \text{ kW}$$

Solution (c)
From $Q = u \cdot a$ m³/s, mean velocity of the pumped water

$$u = Q/a = 4 \times 0.16/[\pi \times (0.2)^2]$$
$$= 5.09 \text{ m/s}$$

From the Bernoulli equation

$$Z_1 + P_1/\rho g + u_1^2/2g = Z_2 + P_2/\rho g + u_2^2/2g + \text{losses metres of}$$
$$\text{fluid flowing}$$

Now at the lake, water levels $P_1 = P_2$ and $u_1 = u_2$. Then by rearranging the Bernoulli formula the hydraulic losses dh handled by the pump will be

$$dh = (Z_2 - Z_1) + [k(u^2/2g) + (4fLu^2/2gd)] \text{ metres of water flowing}$$
$$= (\text{static lift}) + [\text{frictional losses in the pipe}]$$

Note that with the bellmouth entry and exit connections, shock losses are taken as negligible. Substituting

$$dh = 150 + 1.1[(5.09)^2/2g] + [4 \times 0.005 \times 180 \times (5.09)^2]/(2g \times 0.2)$$
$$= 150 + 1.45 + 23.77 = 175.22 \text{ m of water flowing}$$

and $dP = dh \cdot \rho \cdot g = 175.22 \times 1000 \times 9.81 = 1\,718\,908$ Pa

$$= 17.2 \text{ bar gauge}$$

The volume flow of water through the pump is given as 0.16 m³/s so the mass transfer of water is 160 kg/s and the pump duty will be 160 kg/s at 17.2 bar gauge.

Solution (d)
The theoretical power required to drive the pump $P_w = Q \cdot dP$ Watts.

$$P_w = 0.16 \times 1\,720\,000 = 275\,200 \text{ W} \quad \text{so} \quad P_w = 275 \text{ kW}$$

Practical power to drive the pump = 275/0.7 = 393 kW this would be drawn from the Grid during the off peak period.

Solution (e)
The total volume handled by the pump = $0.16 \times 7 \times 3600 = 4032$ m^3. Let y = the operating hours for the turbine, then the total volume handled by the turbine = $0.39 \times y \times 3600 = 4032$ from which $y = 2.87$ h.

SUMMARY FOR EXAMPLE A3.4
It is assumed that the pump is located in parallel with the turbine and the hydraulic loss in the suction pipe to the pump from the lower lake is ignored.

A3.4 Marine turbines

The potential for using the power of tidal streams to generate electricity for homes and businesses is considerable. There are numerous places around the shores of the United Kingdom where the technology could be applied to provide a virtually continuous supply.

A considerable advantage with the use of marine turbines is that the superstructure above sea level, and therefore that which can be seen, is minimal. There is also no harm to marine life as the blades rotate slowly.

An array of turbines comprises a series of towers each supporting two turbine rotor sizes of 15–18 m diameter that look much like those used for wind turbines. The generators are housed in the top of the towers. The theory for the calculation of power and energy generation is similar to that for wind turbines. However, whereas wind is multidirectional the tidal stream is bi-directional and tends to be like a river of fast moving water flowing through surrounding less fast moving water. Water density has a significant influence in marine power calculation. The effect is offset somewhat with the mean site specific velocity which is not so high as it is for the mean wind speed in wind turbine calculation. On energetic sites, the mean spring peak (MSP) velocity is around 3 m/s. These sites have to be rigorously investigated to ensure, among other matters, maximum use of the tidal stream and suitable power connection to the mainland.

As indicated above, the theoretical power equation for wind turbines applies here. Theoretical power equation (A3.1) $P_w = 1/2 \cdot a_s \cdot u^3 \cdot \rho$ Watts. Since water density is quite different to air density, equation A3.1 must account for this. Equation A3.2, which calculates the theoretical annual (H) energy in kWh, must be revisited.

Energy (H) in kWh can be evaluated using P_w in Watts. Thus theoretical $H = P_w \times (8736/1000)$ kWh where there are 8736 h in 1 year and there are 1000 W in 1 kW. Substituting for P_w we have

$$H = \left(\tfrac{1}{2} \cdot a_s \cdot u^3 \rho\right)(8736/1000)$$

Taking water density = 1000 kg/m³, then theoretical energy for marine turbines

$$H = (4368 a_s \cdot u^3) \text{ kWh/annum} \tag{A3.4}$$

There are three efficiencies that apply to marine turbines:

η_g = generator efficiency which is around 0.95;

η_r = rotor efficiency which is in the region of 0.45;

η_u = efficiency of utilisation that is in the region of 0.85.

Efficiencies (η_g) and (η_r) apply to the calculation of potential turbine power (P_w) and efficiencies (η_g), (η_r) and (η_u) apply to the determination of potential annual energy output (H).

There now follows an example of potential output from an array of marine turbines.

Example A3.5 Find the potential power output and annual energy output from an array of marine turbines

An array of six tidal turbines is located in an estuary where the mean spring peak (MSP) velocity is 3 m/s. Two 18 m diameter rotors are mounted on each of the three support towers. The site specific rated velocity is expected to be 40% of MSP velocity. Adopting the appropriate efficiencies above, determine the potential power output of the array and the estimated annual energy output.

Solution
From equation A3.1 $P_w = 1/2 \cdot a_s \cdot u^3 \cdot \rho$ (W). The site specific velocity $u = 3 \times 0.4 = 1.2$ m/s and water density is 1000 kg/m³. Substituting theoretical power

$$P_w = 1/2[\pi(18)^2/4] \times (1.2)^3 \times 1000 = 219\,861 \text{ W each}$$

practical power

$$P_w = 219\,861 \times 0.95 \times 0.45 = 93\,990 \text{ W each}$$

practical power output from six rotors = $93\,990 \times 6 = 563\,940$ W = 564 kW

From equation A3.4, $H = 4368 \cdot a_s \cdot u^3$ kWh/annum. Substituting, theoretical energy

$$H = 4368 \times [\pi(18)^2/4] \times (1.2)^3$$
$$= 1\,920\,707 \text{ W each}$$

practical energy

$$H = 1\,920\,707 \times 0.95 \times 0.45 \times 0.85$$
$$= 697\,937 \text{ kWh/annum each}$$

Practical energy from six rotors $= 697\,937 \times 6$
$$= 4\,187\,623 \text{ kWh/annum}$$

SUMMARY FOR EXAMPLE A.3.5
Potential output from the array is 564 kW and the estimated annual energy from the array is 4.18 GWh.

A3.5 Solar irradiation and the solar constant

The fraction of the Sun's energy reaching the outer atmosphere of the Earth may be calculated approximately if it is assumed that the Earth travels in a circular path around the Sun. The fraction will be the ratio of the Earth's disc area to that of the spherical surface area described by its radial path around the Sun. The surface area so described will receive all the Sun's radiation whereas the Earth's disc will receive the fraction calculated. Thus the fraction = (area of Earth's disc)/(surface area of described sphere). Refer to Figure A3.6.

The Earth's radius is 6436 km and the radius of the Earth's path described around the Sun is (150.6×10^6) km.

$$\text{The fraction} = \pi(6436)^2/4\pi(150.6 \times 10^6)^2 = 4.56 \times 10^{-10}$$

Adapting equation 4.2, the Sun's total emission $Q = \sigma(T^4)A$ W. The Sun's temperature is 6000 K and its radius is (6985×10^5) km. The total emission

$$Q = 5.67 \times 10^{-8}(6000)^4 \times 4\pi(6985 \times 10^5)^2$$
$$= 4.505 \times 10^{26} \text{ W}$$
$$= 4.505 \times 10^{23} \text{ kW}$$

That reaching the Earth $= 4.505 \times 10^{23} \times$ (fraction received)
$$= 4.505 \times 10^{23} \times 4.56 \times 10^{-10}$$
$$= 2.0543 \times 10^{14} \text{ kW}$$

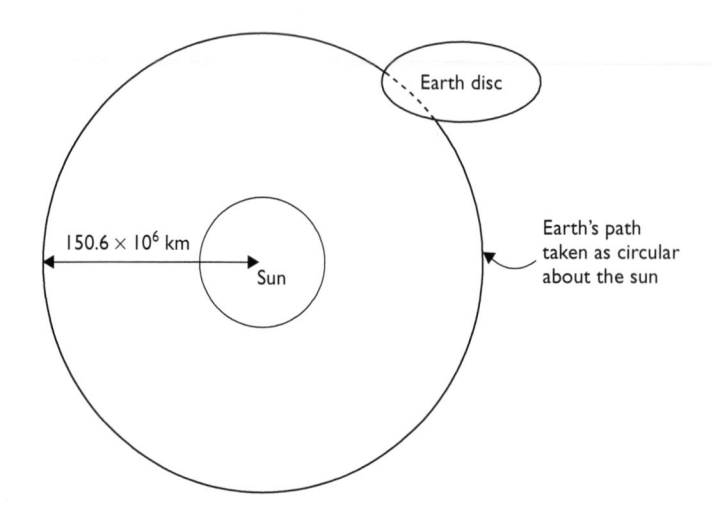

Figure A3.6 Sun and Earth as a disc. The Earth's motion around the Sun considered circular. Fraction of Sun's radiation received by the Earth = (area of Earth's disc)/(surface area of described sphere).

The solar constant

$$I = Q/A = 2.0543 \times 10^{14}/\text{disc area kW/m}^2$$
$$= 2.0543 \times 10^{14}/\pi(6436 \times 10^3)^2$$
$$= 1.5786 \text{ kW/m}^2$$

This compares with the measured solar constant perpendicular with the Sun's rays outside the Earth's atmosphere of $I = 1.388$ kW/m^2. The discrepancy is accounted for in the assumptions made above.

The solar intensity on a horizontal surface at latitude 51.7 on 21 June at 12.00 is: $I = 0.85$ kW/m^2. The solar intensity on a vertical surface facing south at latitude 51.7 on 22 September and 22 March at 12.00 is: $I = 0.700$ kW/m^2. The solar intensities given here are the maximum values in the year for sky clarity of 0.95, cloudiness factor 0.0, ground reflectance factor 0.2 and altitude 0–300 m.

Solar thermal collectors

There are a variety of solar thermal collectors in use for water heating. They are of the fixed type and therefore need to be located south facing and tilted to attract as much solar radiation as possible. The angle of tilt is often fixed by the slope of the roof but it is more important than the orientation.

The most common types of solar thermal systems are used for hot water services and swimming pools. However, they can also be used to heat a suitably located thermal store during summer, and can supply a space heating system during the winter season.

Since a typical non-reflective surface has a high absorptivity, from Kirchoff's law, it will also have a high emissivity. A solar thermal collector should have a selective surface that allows high absorptivity but low emissivity. Refer to Example A3.6. If greenhouse type glass is used to protect the collector surface it will allow the passage inwards of short wave solar radiation and assist in preventing the transmission of long wave radiation from the collector surface outwards through the glass.

Collector types

There are two types of solar thermal collector:

1   The flat plate collector that commonly includes an aluminium/copper absorber with copper tubes clamped to it in serpentine form or parallel format. Stainless steel is used for directly fed panels serving swimming pools. The thermal bond between the tube and absorber backing is essential for good heat transfer by conduction. The following collectors are more efficient but less robust.
2   The direct flow vacuum tube collector consists of a series of evacuated glass tubes, connected to a header, each having a parabolic reflector. The primary fluid flows in a copper tube located within the evacuated tubes. The tubes can be rotated individually for alignment with the sun.
3   The heat pipe vacuum tube collector has some similar characteristics. However the heat transfer medium operates on the heat pipe principle by evaporating as it rises in the copper tube due to irradiation where it condenses in the header and gravitates back down the tube to repeat the process. Thus heat transfer is by evaporation and condensation. Unlike the foregoing, for this collector to operate, it must be installed with a minimum inclination to effect the heat pipe principle.

Collection ability

Clearly the collecting ability of the solar thermal panel relies on a number of factors:

* Latitude of collector location.
* Seasonal solar irradiation intensity, sky clarity, ground reflectance and location altitude.
* Geometric relationship between the panel and the path of the sun.
* The ability of the collector to respond to diffuse irradiation.

Example A3.6 Find the rate of energy collection from solar thermal collectors and the collection efficiency

A flat plate solar collector has a selective surface with an absorptivity of 0.95 and an emissivity of 0.05. This is achieved using a special coating applied to the collector. The coefficient of convective heat transfer is 3 W/m²K at the collector surface. If the area of the collector is 2 m² and there are four connected to the same system, calculate the rate of energy collection and the collection efficiency at a time when the irradiation is 820 W/m² Take the collector temperature as 65°C and outdoor air temperature as 27°C.

Solution
The convection loss from the outer surface of the collector adopting equation 3.16, $Q = h_c \cdot A \cdot dt = 3 \times (2 \times 4) \times (65 - 27) = 912$ W. Adapting equation 4.12, the radiation loss $Q_r = \sigma \cdot e \cdot T^4 \cdot A$ Watts. So

$$Q_r = 5.67 \times 10^{-8} \times 0.05 \times (273 + 65)^4 \times (2 \times 4) = 296 \text{ W}$$

The net rate of collection $= (0.95 \times 820 \times 2 \times 4) - 912 - 296$
$$= 5024 \text{ W}$$

Solar irradiation $= 820 \times (2 \times 4) = 6560$ W

The collection efficiency = net collection/incident irradiation
$$= (5024/6560) \times 100 = 76.6\%$$

Note: This is the collection efficiency and not the overall efficiency of the solar collector and storage system.

Example A3.7 Find the rate of energy absorption from solar thermal collectors

Calculate the rate of energy absorption on a flat plat collector having an area of 3 m² and positioned normally to the Sun's rays. The surface temperature of the collector is 70°C and outdoor air temperature 21°C. Its absorptivity to solar radiation is 0.95 and the emissivity of the plate is 0.05. The convective heat transfer coefficient at the collector surface is 3 W/m²K. Take the solar constant as 1.388 kW/m² and the transmissivity of the upper atmosphere as 0.63.

Solution

$$\text{Rate of collection} = 0.95 \times 1388 \times 0.63 = 831 \text{ W/m}^2$$

$$\text{Radiant loss} = 5.67 \times 10^{-8} \times 0.05(273 + 70)^4$$
$$= 39.24 \text{ W/m}^2$$

$$\text{Convection loss} = 3 \times (70 - 21) = 147 \text{ W/m}^2$$

$$\text{Net rate of absorption} = 831 - 39 - 147 = 645 \text{ W/m}^2$$

$$\text{Net rate of energy absorption} = 645 \times 3 = 1935 \text{ W}$$

SUMMARY FOR EXAMPLES A3.6 AND A3.7

In both the above examples the net rate of solar irradiation collected is then transferred to the collecting medium which is usually water treated with an antifreeze agent. There is a loss of efficiency here and also at the point where this heated water imparts its energy at the heat exchanger to the water used for consumption. An overall efficiency of about 50% is expected from the use of current solar thermal collectors. Refer to A4.7.

The solar collectors described above are of the flat plate type through which the primary water flows and is heated as it passes along the tube coil in the panel.

A3.6 Photovoltaics

The photovoltaic (PV) effect occurs when streams of sunlight called photons irradiates a PV cell. The effect is to free electrons present in the cell and move them to an attached wire. During the process, they carry with them an electrical charge.

As the charge is small, even in strong sunlight, several cells are interconnected to form modules. Groups of modules are mounted together to form a panel and the panels are interconnected to form a PV array. Modules can be integrated into standard facade cladding systems and take the place of various types of roof tile. They can also form part of external moveable solar shading screens.

There are various types of cells used in PV modules. The most widely used are single crystal silicon, multicrystal silicon and amorphous silicon, which is used indoors as it responds better to fluorescent light. Other types of PV cells include polycrystalline cells, ribbons of polycrystalline silicon and thin film solar cells.

Silicon, an abundant natural resource, easily absorbs photons and provides electrons. To assist this natural process, silicon cells are coated with boron and phosphorous. A typical cell consists of a wafer thin layer of phosphorous-doped silicon in close contact with a layer of boron-doped silicon. This chemical treatment creates a permanent electric field. The electricity generated is direct current (DC). Recent advances in photovoltaics means that DC electricity can be generated from a clouded sky.

Unlike the ingot growth techniques used for traditional crystalline silicon PV cells, thin film cells can be manufactured in larger sizes using automated continuous production processes. Thin film cells can be deposited on low cost flexible substrate materials such as glass, stainless steel or plastic in virtually any shape, giving it potential for use in many building applications. Thin film cells have many advantages over their thick film counterparts as they use much less material and can be made as large single units.

PV systems

There are two main types of PV system: storage and grid connected.

1. The storage system allows the electricity generated during the day to be stored for use then and at night time. The method of storage is usually by battery. The storage system for a PV array will provide a DC current, so electrical appliances must be compatible. With the storage facility, the system can be independent of the National Grid and is clearly suitable where there is no available mains electricity.

    If the appliances to be connected to a PV array are compatible with AC current a DC/AC inverter is required.
2. Where mains electricity is available, the PV system can be connected to it via a DC/AC inverter and a two-way meter. This allows the PV array to contribute to the demand for electricity during the hours of daylight.

The use of PV systems is now well established with arrays located on vertical and inclined facades and on building roofs. Domestic use is possible under micro-generation technologies that include the complete solar roof that replaces conventional roofing products with a combination of solar electric PV tiles and solar thermal panels for generating on site electricity and hot water needs. The combined solar thermal and PV panels, although presently expensive, may be useful in tight urban sites where the roof area for solar energy is at a premium.

There are now numerous examples of the use of PV systems in commercial and industrial applications. The take up in domestic use is at present low due to cost. However, this may change significantly as cost comes down.

The nominal peak power from a module of 36 series connected monocrystalline silicon cells (as opposed to thin film cells) is 90 W at 18 V and 5 A. The size of the module is approximately 1200 mm by 530 mm by 44 mm depth, and it weighs between 5.5 kg and 7.5 kg.

A3.7 Biomass

The use of the term biomass can be broken down into two main categories:

> Woody biomass includes thinnings, and trimmings from woodlands, untreated wood products, willow, short rotation coppice (SRC), straw and miscanthus.

Non-woody biomass includes animal wastes, industrial and biodegradable municipal products from food processing and high energy crops such as rape, sugar cane and maize.

If biomass is allowed to rot naturally it releases methane which passes into the atmosphere and is up to 21 times as harmful as CO_2. The term biomass is used for fuels that derive from natural vegetation, which in the life cycle fixes carbon dioxide from the air during the process of photosynthesis. Thus when the vegetation is harvested and prepared for combustion to produce heat, the amount of carbon dioxide released is only slightly more than that absorbed during the growing cycle. (The harvesting, preparation and transport of the biomass incurs a carbon penalty.) Thus biomass is considered as carbon neutral with CO_2 emissions of 0.035–0.06 kg/kWh compared with 0.19 kg/kWh for natural gas and 0.27 kg/kWh for oil.

Wood in the form of sized and split logs has been used domestically for centuries as a means of heating and cooking and in the forests of Europe and elsewhere it is still used in this way in rural communities.

The calorific value of biomass varies from 8 kJ/kg to 20 kJ/kg. This compares with an average for coal of 27.4 kJ/kg and oil at 40 kJ/kg. One of the uses of wood products in the commercial and industrial sectors in CHP and centralised power generation is in the form of pellets or chips prepared directly from bespoke managed woodland or from wood waste from saw mills and wood machining processes. Moisture must be evaporated from the fuel before it can be used. Rural communities using logs as a means of heating and cooking lay up newly timber after splitting for at least a year to allow excess moisture to depart.

Pellets have the following advantages over other types of wood fuel:

- Less volume to transport and store with consequent fewer deliveries.
- Consistent size and moisture content.
- Versatility – can be used in domestic boilers and commercial/industrial boilers.
- Less ash and emissions.
- Can be delivered to the boiler combustion chamber easily.
- Easy to ignite.

In the south of the United Kingdom miscanthus is successfully grown as a biomass fuel. Compared with SRC, when harvested, it is comparatively dry and the yield potential is also higher at 10–20 tonne per hectare compared with about 10 tonne per hectare for SRC. Some boiler manufacturers have responded positively to this renewable energy source and storage, feed and combustion systems are now well advanced and highly controllable.

A modern wood burning boiler can operate at efficiencies of 90% with carbon monoxide emissions as low as 100 mg/m^3. It produces higher flue gas temperatures than a boiler fired by natural gas and operates best at continuous load. It provides a very good replacement for coal fired boiler plant. As with oil fired and coal fired boilers storage on site is required. A building with a heating design load of 50 kW is likely to burn 8500 L (8.5 m^3) of oil per annum. A boiler burning wood pellets would require 17 tonnes per annum. This would occupy a space of 26 m^3. Clearly, it is not necessary to provide a years' storage on site but the comparison with oil is valid.

Delivery of the fuel in pellet form only is by blower through a 100 mm diameter pipe from the delivery lorry. Ro-bin delivery comes in the form of a skip or bin which replaces the empty bin on site. This requires the site to have at least two bins to ensure continuity of fuel to the boiler plant. Tipper delivery simply delivers the fuel into an accessible store.

In applications where load variations are substantial, the use of a buffer tank is considered. This provides a volume of water as a heat store between the heat supply side and the load side of the heating system. The size of the buffer tank should be about 15 litres per kilowatt of boiler rating. The boiler is controlled from the mean temperature of the water in the buffer tank. As the mean temperature rises so the boiler output is reduced to a point where the boiler shuts down and its residual heat is stored in the water contained in the buffer tank.

Biomass fuels are increasingly being used in CHP plant by passing them through a gasifier in the same way as coal gas was produced in the last century.

Biofuels

Biofuels such as biodiesel and bioethanol are at present blended with diesel and petrol although they can be used on their own in internal combustion engines when they are also considered as carbon neutral.

Recent research has been undertaken into exploring the possibility of using biodiesel as a replacement for kerosene in domestic oil fired boiler plant. It is sourced from used vegetable oil (fatty acid methyl ester – FAME) and harvested rapeseed (rapeseed acid methyl – RME). A property comparison with kerosene and gas oil is given in Table A3.1.

Table A3.1 Properties of biodiesel and mineral oils

Property	Units	Biodiesel	Kerosene	Gas oil
Density	kg/m^3	860–900	795–805	840–875
Viscosity at 40°C	(mm)^2/s	3.5–5.0	1.0–2.0	2.5–4.5
Flash point	°C	>101	>39	>62
Net CV	MJ/kg	37	43	42.3

From Table A3.1 biodiesel shows a good comparison with the calorific value but the flash point is substantially higher leading to the need for pre-heating prior to combustion, if biodiesel is used on its own without mixing with mineral oils.

In 2003, UK CO_2 emissions from domestic oil fired plant were around 7.5 million tonnes. If RME B20 biodiesel is used in the domestic market – that is to say – a mix of 20% with mineral oil, the CO_2 emissions saved would be 1.5 million tonnes per annum and the land required for rapeseed would be in the region of 330 000 ha.

A3.8 Combined heat and power

Currently CHP does not commonly derive from renewable sources of energy. The exceptions are one or two small power stations that also provide space heating local to the generating facilities and operate from wood as the combustible material. The second exception is CHP from fuel cells where the hydrogen is derived from renewables.

It is commonly known now that power-only generation suffers from very low overall efficiency (around 35%). This is now unacceptable whether fossil fuel is used with a 65% waste of a finite resource or indeed whether a renewable energy is used with the same loss from the renewable resource.

A key benefit of CHP is its efficient utilisation of precious fuel resources. Among its benefits are

- an overall reduction in energy costs;
- increased fuel conversion efficiency in comparison with conventional alternatives;
- reduction in carbon emissions, due to efficiency levels and/or resulting from the use of renewable fuels or fossil fuels with a lower carbon content;
- security of supply.

Combined heat and power can provide the space heating and some of the electrical needs of the site from the use of one fuel. The heat produced during the process of generating electricity is used for space heating and hot water services. The effect is to raise the plant efficiency from 35% for a power-only plant to as high as 80%, thus making much better use of the fuel resource.

The applications of CHP include commercial and industrial sites, community heating and district heating. If absorption cooling is added to the CHP plant we have electricity production, space heating and air conditioning available for local use. This is known as tri-generation. The fuels in use at present are natural gas and oil both of which are, of course, fossil fuels but they are being used to generate heat, power and possibly air conditioning in one combustion process and for the time being they are accepted as contributing to reductions in the emission of carbon dioxide.

CHP plant

Modern CHP is delivered to site in the form of packaged equipment containing the engine, heat exchanger, electrical generator, system integration connections, exhaust system and controls. For the time being the fuel commonly used is either diesel or natural gas. The prime movers include the following:

Reciprocating engines. These include the spark ignition engine for gaseous fuels such as natural gas or renewable fuel such as biogas and compression ignition engines using diesel or a biodiesel mix. More information is given in another publication in the series – see Section A3.10.

Micro-engines/turbines. Package CHP using the micro-engine and turbine technology largely for the domestic market (micro-CHP).

Stirling engine. The external combustion engine patented by Dr Robert Stirling in 1816. It is now developed for the domestic market. It can provide thermal and electrical energy with low levels of noise and air pollution as micro-CHP.

Fuel cell. Fuel cells convert hydrogen directly into electricity and heat energy. These are discussed in Section A3.9.

Assessment of CHP in buildings

Individual buildings have unique energy demand patterns. To realise the benefits of CHP the following require consideration:

- The accurate determination of the building's thermal and electrical energy profile and heat to power ratio.
- Selection of appropriate CHP plant.
- The integration of CHP with the building services – existing or proposed.
- Assessment of fuel and grid electricity tariffs and costs.
- Feasibility using grid electricity and supply locally generated excess power to the grid.

Combined heat and power is best utilised when supplying constant electrical and thermal loads, and, of course, for most buildings these fluctuate through the seasons of the year. It may be appropriate to consider CHP for the base loads, relying on the National Grid for peaks and a thermal store that is brought onstream in the summer for topping up the space heating in the winter. Absorption cooling will increase the heat demand in the summer when space heating is not required. Thus a site that has inconsistent demands for thermal and electrical energy may be deemed not suitable or may need to incorporate additional design features to improve the viability of CHP.

Heat to power ratios

Variation of thermal and electrical demands mean that building heat to power ratios H/P are transient and therefore it is unwise to try matching this with the output ratio of just one CHP unit. A selection of potential CHP units need to be appraised and consideration given to installing more than one unit

Micro-CHP package units are produced up to about 200 kW with heat to power ratios of 2 to 3. The power generators can be linked to the electric grid so that power may be purchased if demand exceeds capacity, or sold to the grid if demand falls.

It is important to select the CHP plant that matches the building's energy needs for the maximum length of time during the year at optimum plant output. These considerations allow flexibility in the application of CHP to buildings.

CHP with additional features

As mentioned above, in order to get the most out of a CHP plant it must operate at its optimum output for as long as possible during a typical year. In many buildings the summer season is likely to be the time of year when with no space heating and good daylight conditions the plant may be running at low load with the consequent drop in electrical power output. If the electrical output of the plant is maintained, the heat energy generated may have to be stored.

The excess heat can be used by introducing thermal storage to capture the excess heat generated during the summer for use in the winter. If there is a requirement for air cooling the use of absorption refrigeration plant can be employed to use some of the excess heat as well.

This, of course, adds cost to the services systems. However, whole life costing or life costing analysis may well show a better return as fossil fuel prices increase over time. Whole life costing is discussed in another publication in the series. Refer to Section A3.10.

A3.9 Fuel cell CHP

A fuel cell consists of an electrolyte sandwiched between two electrodes. Oxygen or air is passed over the cathode and hydrogen over the anode generating water, electricity and heat. Thus, the fuel cell in this context provides combined heat and power.

Individual fuel cells provide a limited output so they are assembled into modules or stacks to provide the required voltage and current. When supplied by hydrogen, fuel cells have zero CO_2 emissions. However, a variety of fossil fuels can be used while a hydrogen infrastructure is put in place. The operation of a fuel cell in this way will incur emissions into the atmosphere but this is much lower than the emissions from the combustion process of the cleanest fossil fuel.

The current use of fuel cells includes:

- Powering cars, buses, boats and trains
- Hospitals, call centres, police stations and banks
- Waste water treatment plants
- Landfill sites producing methane can use it for fuel cells to produce electricity
- Community heating and electricity
- Domestic properties
- Miniature fuel cells are in the pipeline for cell phones, laptop computers and portable electronic equipment.

Types of fuel cell

There are several different types of fuel cells. They are usually classified by the type of electrolyte they use. Some types are suitable for stationary power plants, others for cars and portable applications. The following types of fuel cells are the focus of current development work.

Proton exchange membrane fuel cell (PEMFC). This type of fuel cell is well advanced and will probably end up powering vehicles and buildings.

Solid oxide fuel cells (SOFCs). This type of fuel cell will be used for large scale stationary CHP, such as factories and towns. High operating temperatures of around 1000°C will allow steam to be produced to drive turbines thus increasing the electricity output.

Solid polymer fuel cells (SPFCs). Very promising for both vehicles and stationary applications.

Phosphoric acid fuel cells (PAFCs). This fuel cell is more developed than the SOFC and SPFC. It has a higher temperature than the PEM fuel cells and like the SOFC has a longer warm up time.

Alkaline fuel cells (AFCs). This is a simple device and has applications in vehicles. However, it is susceptible to contamination and requires pure hydrogen and oxygen.

Molten carbonate fuel cells (MCFCs). This type is suitable for stationary power and CHP applications and has been demonstrated at 2 MW power output. It operates at 600°C so does not require the high specification materials required by the SOFC, although the temperature is clearly sufficient to produce steam for extra power generation.

Fuel cell efficiency

Fuel cells have a heat to power ratio of 1 : 1 with overall efficiencies of 80% when fired on hydrogen. If a reformer is added to convert other fuels to hydrogen the efficiency is significantly reduced, but it can be used to advantage as an interim measure. Fuel cell systems can maintain high efficiencies at loads down to 50% which makes them attractive for installation in buildings which often operate at low loads.

The fuel cell has a number of advantages over heat and power derived from combustion engines. Efficiencies can be higher and CO_2 emissions reduced. They also produce lower emissions of oxides of nitrogen, hydrocarbons and particulates. Fuel cells running on hydrogen derived from renewables will emit only water vapour.

These systems also tend to have lower noise levels than conventional alternatives. The life expectancy is 20 years although the fuel cell stack requires replacing every 5 years. Hydrogen is a limitless fuel that can be produced from electricity and water and a hydrogen economy promises to eliminate many of the problems that the fossil fuel economy creates.

In the meantime, natural gas will be used in the fuel cell and the next generation of combi boilers for the domestic market will generate the heat, hot water and electricity base load for the typical home – thus, reducing the carbon dioxide emissions by about 3 tonnes a year.

A3.10 References and further reading

Heating and Water Services Design in Buildings, 2nd edition, K.J. Moss, Taylor & Francis, 2003.
CHP, Lee Hargreaves, *EIBI*, January, 2004, CPD Collection.
Solar Thermal, Eric Hawkins, *HPM*, September, 2004.
Solar Thermal, Tim Dwyer, *CIBSE BSJ*, September, 2004, CPD Collection.
Designing for Biomass, Gavin Gulliver-Goodall, *EIBI*, January, 2005.
Biomass, Tim Dwyer, *CIBSE BSJ*, September, 2005, CPD Collection.
Fuel Cell CHP, Phil Jones, *EIBI*, December, 2005, CPD Collection.
Solar Thermal, Tim Dwyer and Bill Sinclair, *CIBSE BSJ*, February, 2006, CPD Collection.
The Future in Numbers: 50 Technologies, *CIBSE BSJ*, Special Supplement, August, 2006.
Photovoltaics, Trevor J. Price, *EIBI*, November/December, 2006, CPD Collection.
Marine Turbines, Peter Fraenkel, Marine Current Turbines Ltd.

Biomass Boilers, R. Bradford, N. Monether, G. Israel, *EIBI*, January, 2007, pp. 36–41.

A3.11 Appendix closure

This appendix is an introduction to some of the renewable technologies. Local authorities in the United Kingdom are now calling for a proportion of energy consumption in new and renovated buildings to be offset using on site or local renewable energy sources. This clearly requires building designers to have specialist knowledge of renewable energy systems that can be used on or near the site.

Hydro power, marine turbines and wind turbines used on wind farms are considered as green electricity supplementing that generated by traditional power stations for the National Grid. Green electricity, however, is also generated to supply local communities with the excess electricity passing into the grid.

The current emphasis is to decentralise generation capacity. However, this is a political issue as is the future supply of natural gas from outside the UK.

Appendix 4

Towards sustainable building engineering

A4.1 Introduction

Fifty years ago central heating was introduced to the domestic market. One hundred years ago and more potable water, gas and electricity were piped to many homes in the United Kingdom. These services are now instantly available at the space heating time switch, the light switch and the tap. We in the West have grown up with them and what were luxuries are now necessities. We do not even think about the time switch, switching the lights on or for that matter turning the tap on. The time settings are pre-planned and the responses are instant. For example, when we enter a room many of us turn the lights on whether they are needed or not. It is not necessary now in daily routine to prepare and replenish the stove, order the day to suit daylight hours or walk to the pump or well for water.

Fossil fuel has made life easy and the going good. This presents us with serious challenges as we in the West are now called upon to think about these actions that are taken for granted. There are two reasons for this and they are linked together; first the planet's reserves of raw materials and fossil fuel are finite and the world's supply of fresh water, although constant, is increasingly limited due to the rapid rise in the world population. The domestic consumption of water in the West has also doubled in the last 50 years. The West and the emerging countries that are industrialising are following the model of economic growth year on year. This is exponential growth that the laws of thermodynamics tell us is unsustainable. If, for example, the growth rate is at a steady 3% per annum, in 23 years it would have doubled. In the West economic growth has been sustained at varying rates per annum since the industrial revolution and 3% might be a good average figure, so every 23 years since then, give or take, the economic growth of the United Kingdom has doubled.

The second reason which, of course, is a direct result of the first is down to climate change that is seriously being accelerated from harmful gases released into the atmosphere as well as the pollution caused by waste.

It is not being suggested that a return to 100 years ago is needed but new ways must be found of responding to the challenge that the planet cannot replenish at the rate man is consuming its bounty or recycle the pollution this is causing.

Fifty years ago we were largely unaware of the planet's waning health. Now most enlightened people realise that Western societies must in time convert to the principles of re-use, recycle and reduce – in other words, to move to sustainable societies. This might seem an impossible task but we live in one world and its biophysical systems cannot be changed or accelerated. Biophysical sustainability of the planet – that is, its natural replenishment governed by photosynthesis and the laws of thermodynamics – is not only non-negotiable it is preconditional to any manmade concept of governance.

Clearly the built environment whether it is on the drawing board, new, renovated or existing impinges directly on this issue and in each of our disciplines we have to seriously respond to reducing energy from fossil fuels and the dumping of waste.

A4.2 Thermodynamics and sustainability

Seven years ago Sir Jonathon Porritt, Chairman of the Sustainable Development Commission, gave an address to the Institution of Incorporated Engineers that focused on engineering in general and sustainable development relevant to building engineering.

His seven basic principles for what he called eco-efficiency were and still are timely and thought provoking:

- Reduce the material intensity of products.
- Reduce energy intensity.
- Reduce toxic dispersal.
- Enhance material recyclability.
- Maximise the sustainable use of resources, such as energy and raw materials.
- Extend the durability of the product.
- Increase maintainability so that the use of the product is increased for a lower resource input.

The three markers above of re-use, recycle and reduce are evident within these principles and apply to individual products used in the building and building services industries as well as the buildings and service systems they contain.

The difficulty facing the industrialised nations is that we have got used to extracting and processing raw materials at a greater rate than nature can replenish them.

Sir Jonathon believes that scientists and environmentalists are now beginning to converge around an understanding of what the basic scientific principles relating to the activities of man on his world are:

- The conservation of energy and matter
- Energy and matter tend to disperse over time
- Energy and matter cannot disappear or be created
- Increases in net material quality are almost exclusively derived through photosynthesis.

The reader will see the principles of thermodynamics and cell biology here.

- The first scientific principle means that waste material and the products of combustion from fossil fuels do not disappear and that the entire concept of waste disposal and the disposal of combustion products is an illusion. Approximately 100 tonnes of raw material along with the energy from fossil fuels enters the industrial process to generate 1 tonne of product.
- The second scientific principle relates to the dispersal of energy and matter. Energy derived from fossil fuels produces products of combustion that disperse, with consequences, into the atmosphere. Natural resources that are mined and extracted and end up as spoil eventually disperse back into nature. For example, the product steel eventually rusts. Neither the products of combustion nor rust can return to fossil fuel or steel. There are many products that do not disperse or degrade, of course. Glass and some plastics are examples and here the markers reduce, re-use and recycle must apply.
- The third principle is that matter and energy cannot disappear. What society consumes is not the raw material itself but the products made from it. If society consumes our natural resources faster than the Earth can supply them, it is obviously becoming poorer and the serious business of dealing with the products of combustion and 'unnatural' waste matter will not go away.
- The fourth principle is one of cell biology and asserts that increases in net material quality are almost exclusively derived through sun-driven processes – essentially through photosynthesis. It is the open window for the planet making it, in thermodynamic parlance, an 'open system'. This is the prime mover of the planet's replenishment; effectively lowering the planet's entropy (see the third law). Photosynthesis is a chemical process occurring in green plants, algae and many bacteria by which water and carbon dioxide are converted into food and oxygen, using energy absorbed from solar radiation. The reactions take place in the chloroplasts which is the microscopic structure within the plant cell. Molecules of the light absorbing pigment chlorophyll are embedded in the cell membranes. During the first part of the process, light is absorbed by the chlorophyll and splits water into hydrogen and oxygen. The hydrogen

attaches to a carrier molecule and the oxygen is set free. The hydrogen and light energy build a supply of cellular chemical energy, adenosine triphosphate (ATP). Hydrogen and ATP convert carbon dioxide into sugars including glucose and starch.

Forests, for example, therefore rely on carbon dioxide in the atmosphere and rain to survive and in the chemical process that takes place release oxygen back into the atmosphere. Wood used as biomass for fuel is, therefore, considered as carbon neutral. If climate change has the effect of reducing or eliminating rainfall on areas of forest, the trees will suffer or die and the absorption of atmospheric carbon dioxide and the consequent release of oxygen will suffer or be lost.

A4.3 The laws of thermodynamics

The laws of thermodynamics relate to heat and work transfer and are hypotheses resulting from observations of the natural world in which humankind live. To date, these laws have not been challenged although they have been interpreted and extended by at least five scientists in the nineteenth and twentieth centuries. *It is important to stress here that the thermodynamic laws relate to the natural world. They attempt to understand what happens in Nature. Scientists and engineers have used the laws to enable an understanding of the processes of heat and work transfer in manmade processes, cycles and devices. But they remain the natural laws of the Planet's life cycle and in this context they apply to the Earth's climate and how it is affected by man.*

Furthermore, the laws overlap because they are natural laws so that a statement attributed to the first law by one authority is found in the second law by another. This should not confuse or diminish the authenticity of the thermodynamic laws as they should be taken together as a tried and tested statement about the world we live in. They are not the only laws that govern the natural world in which we live and a holistic approach is required to engage in an over arching discussion of the state of the planet. However, they do impinge on heat and work energy transfers, an area in which we as building engineers can and must have an understanding and influence.

The zeroth law of thermodynamics

The laws of thermodynamics are about thermodynamic behaviour. They are the planet's natural laws that have as yet not been contradicted and are based upon observable phenomena. They directly relate to the discussion above. This law is concerned with thermal equilibrium.

It states that:

> If a number of discrete bodies are in thermal equilibrium they must all be at the same temperature. Nature is constantly attempting to reach thermal equilibrium.

The first law

This law shows that there is a relationship between heat transfer Q and work transfer W. Such that:

$$Q = W$$

Work can be converted entirely into heat but the reverse process is not possible since some heat will be rejected. Thus:

$$\sum Q \neq \sum W$$

Thus if the heat and work transfers are not equal, any energy difference must have been added to the substance or have been lost from the substance.

This introduces the concept of internal energy U, that is, energy residing within the substance. Thus:

$$\sum Q = dU + W$$

Thus the law introduces the idea of internal energy within a substance and also introduces the concept of the conservation of energy. Bernoulli's equation for frictionless flow in Section 5.3 and frictional flow in Section 7.3 includes examples of the conservation of energy.

Perhaps, the most important feature of the first law in the context of sustainability is that energy can neither be created nor destroyed as it is changed from one form to another (to heat, light, motion, etc.).

The second law

The second law of thermodynamics is concerned with direction. For example, heat flow will occur between two substances of its own accord down a temperature gradient following the natural law. As it flows down it is degraded to the point where heat flow ceases and the two substances are at the same temperature and in equilibrium. Again this law can be observed in nature.

It explains, for example, the natural forces at work in volcanic activity bearing in mind the considerable temperature in the lower and upper mantles of the Earth and the temperature at its surface.

The second law states that the availability of that energy to perform useful work is reduced as it passes through successive transformations.

The third law

This law is concerned with the level of availability of energy in matter to perform useful work. It suggests that the internal energy U of a substance results from random vibration or motion of its molecules and atoms and that the motion is regulated by the temperature of the substance such that at absolute zero (0 K) motion or vibration ceases and the substance forms a collection of perfect crystals.

The term *entropy* is introduced in the third law and shows how it is associated with temperature and with the availability of thermal energy. The second law of thermodynamics tells us that the availability of energy to perform useful work is diminished as it passes through successive transformations. This is sometimes referred to as the law of entropy – *entropy being a measure of the amount of energy in matter no longer capable of further conversions to perform useful work*. Entropy in matter undergoing these transformations within a 'closed system' therefore increases and matter becomes increasingly impotent.

It is only the fact that our planet (an open system) is open to incoming energy (solar irradiation) and a fixed upper limit to concentrations of greenhouse gases in the atmosphere that prevents an inexorable increase in entropy and a decline into chaos. If civilisation continues to grow its economies without thought of the planet's ability to replenish itself as an open system the future is very bleak.

Thus the third law of thermodynamics states:

> At the absolute zero of temperature, the entropy of a perfect crystal of a substance is zero. But at normal temperatures, in closed systems undergoing energy transfers, the entropy in the matter undergoing these energy transfers increases.

Matter can neither be created nor destroyed. Matter does not disappear. Every atom in the universe today has been part of the universe since the Big Bang. So that natural resources that are extracted or harvested to power our economy must eventually return to nature. Products, on the other hand, may be re-used or recycled but not returned to the original raw materials. Examples are numerous: glass, concrete blocks, steel, plastics. Manmade gases, on the other hand, such as carbon dioxide and nitrous oxide may be recycled by nature at work attempting to return to the status quo (equilibrium) but the planet is not winning the battle here. For example, there are now insufficient forests to absorb the minimum amount of CO_2 in the atmosphere so it continues to increase and climate change may move the rain needed by the forests for the process of photosynthesis.

Methane which is a product formed naturally through the process of biodegradation and is the main constituent of natural gas, after it goes through the process of combustion to produce heat, cannot be returned to methane. If methane escapes into the atmosphere it causes considerably more harm than carbon dioxide, its burnt product. Likewise, if a product that is put in landfill is not biodegradable over time, it remains as waste that does not disappear and therefore it pollutes.

A4.4 Power supplies

The National Grid supplies power to commercial sector, industry, the public sector and the domestic sector mostly from power-only generating stations that use fossil fuel and nuclear fuel. These generating stations are located near rivers or the sea for cooling purposes and are usually remote from where the electricity is needed (the second law). In addition to very low thermal efficiency, there are losses over the National Grid network. With the UK government's drive for distributed energy systems, *small-scale heat and power networks* are likely to become more widespread.

The principle of local power generation is not new. Combined heat and power to a residential community, to an industrial site and providing electricity and heat to commercial buildings is well known. The concept of electricity and heat production in the form of micro CHP units in individual homes may well be the next replacement of the combi condensing boiler. Indeed, some multi-storey flats have centralised combined heat and power plants with the power generated to service the lifts, entrances and corridors. The reliance on large power-only generating plants to provide for the base load and peak load is changing in the United Kingdom. The nuclear power stations have contributed to the base load requirements nationally as they cannot be shut down and the power stations operating on fossil fuel have contributed to peak loads.

New building development and the refurbishment of existing buildings now have to fulfill the obligation for generating some of the energy needed on site. Developers are already including renewable energy systems in new and refurbished buildings. This is already having its influence on the building and building services sectors.

A4.5 Products and systems

New buildings and the services within them are now subject to factors that include issues of sustainability and the use of renewable energy systems as well as life cycle analysis.

Of the building stock in the United Kingdom 98% consists of buildings that are more than 5 years old. It is these buildings that will require refurbishment when sold on in the form of thermal upgrade and the installation of renewable energy systems. Many products are now being energy rated. Energy rating certificates are being introduced for new and existing buildings.

Building products

The target for existing housing is to aim for passive buildings. That is, houses that require little or no energy input for space heating. Clearly this is somewhat wishful thinking for the existing housing stock. However, it concentrates the mind on what is the priority here which is thermal insulation and draught control. These two factors should also provide the starting point for existing buildings in the commercial and industrial sector.

Prefabrication of building modules off site reduces errors and waste on site and raises standards. This is now becoming common practice. The thermal mass of a building, particularly the building envelope, is a critical issue for designers looking to keep the building warm in winter and cool in summer. Heavyweight buildings are generally slow to respond to changes in outdoor climate. This can be put to advantage when outdoor temperature drops in winter and rises in summer. The effect of solar heat gains on the indoor climate is delayed for a building constructed from heavyweight materials. Developments in phase change materials that are incorporated into the building envelope could help overcome the disadvantages associated with lightweight building construction.

Phase change materials work by storing thermal energy as latent heat in the process of freezing and melting. A typical example of a phase change material is one that freezes at 16°C and melts at 23°C. Latent heat has a substantially larger capacity to store and release heat than sensible heat. A building normally responds to sensible heat exchange resulting from a rise or fall in temperature.

Environmentally friendly materials are now being used in building construction. They include

- laminated structural beams in wood for supporting roof structures;
- rammed earth walls;
- straw bale walls;
- hemp and lime walls;
- timber frame walls.

Recycled brick, block and stone are also being used for wall construction.

Embodied energy

This is the energy required to process the building product from raw materials. Table A4.1 lists some building products with their density and embodied energy. Table A4.1 emphasises two points in respect of energy conservation.

- The careful selection of building products to minimise the impact on energy resources and on the consumption of mined raw material.

Table A4.1 Embodied energy and density of some building materials

Product	Embodied energy (kWh/kg)	Density (kg/m^3)
Brick	0.83	1700
Dense block	0.50	2300
Medium block	0.42	1500
Light block	1.00	600
Plaster	0.81	600–1300
Plasterboard	1.22	950
Render	0.50	1200
Glass fibre insulation	1.36	12
Timber stud	1.39	500
Glass	3.53	2500
Mild steel	9.44	7800
Aluminium	27.80	1350
Copper	47.20	2700
PVC	22.20	550–650
MDF	3.14	300
Mains water	0.20 kWh/m^3	1000

- At the end of the life of a building, the need to re-use or recycle its component parts as an alternative to resorting to land fill.

The UK construction industry currently consumes 600 million tonnes of material a year, 85% of which is derived from primary resources and only 15% is recycled or re-used.

Building services

Prefabrication of building services modules off site to reduce errors and waste on site and to raise standards is now common practice. Building services systems now require to be designed with energy reduction built in. Boiler plant is now certificated for efficiency. The Seasonal Efficiency of Domestic Boilers in the United Kingdom, the SEDBUK rating, is an example.

Very low temperature space heating systems such as underfloor heating work well with ground/air/water source heat pumps that deliver temperatures around of 45°C. They also allow the condensing boiler to operate at maximum efficiency. Product manufacturers (of circulating pumps, for example) now offer products having low maintenance over a specified life at the end of which the product is recycled by the manufacturer.

Within the next 2 years there will be a new generation of domestic electrical equipment known as *dynamic demand devices* that can sense rising electrical demand via a micro-controller and switch off for seconds at a time without degrading the product or the service it provides. It is not inconceivable that this principle will extend to plant used in building services and will somewhat assist in ironing out the peaks in the National Grid or in locally generated power. Soft start is already present on variable speed drives thus cutting out sudden surges in the power supply.

Long-life fluorescent tubes can now achieve a service life of 70 000 h which is eight times longer than that of the conventional tube. The use of lighting controls, both newly installed and retrofitted, has proved to be acceptable and a valuable contribution to lower energy bills. Micro-generators that convert vibration caused by traffic flow into electricity that can be used for lighting are being developed.

A4.6 The building footprint

Traditionally, the building footprint represents the land area occupied by the building. The concept of the footprint of a building is now understood in two further ways:

- The selection of the site – brown or green field – and its loss to the natural environment. The energy used to prepare the site, supply the materials and construct the building, and at the end of its life the energy required to dismantle, re-use, recycle and restore the site. This is known as its ecological carbon foot print and is the responsibility of the developer or owner.
- The energy and materials used to operate the building during its life. This is called its annual carbon or carbon dioxide foot print and is the responsibility of the building occupier or landlord. Enlightened building owners account for the ecological footprint and the carbon footprint by either planting forests or buying parts of tropical forests to offset the carbon emissions of the building footprints and the natural in-balance caused by building on green and even brown field sites.

A4.7 Scenarios for building services

Two scenarios will be considered here. The first one is using solar thermal collectors that can be retrofitted for domestic use. This and other measures of energy reduction are important in view of the size of the existing housing stock and its contribution to carbon dioxide emissions in the United Kingdom. The second scenario is for new large community housing.

The application of solar thermal collectors

Solar thermal collectors have been introduced in Appendix A3.6. They can be used in conjunction with a conventional boiler that provides space heating and indirect hot water via a storage cylinder having two coils, one for the indirect circuit to the collectors and the other for the primary circuit to the boiler. Alternatively, continuous flow water heaters and some combi boilers can be used in conjunction with a storage cylinder having one coil for the indirect circuit to the solar collectors. This alternative is more efficient and the reader might take time to consider why.

Annual solar irradiation in the United Kingdom varies from 900 kWh/m^2 in northern England to 1100kWh/m^2 in the south. The average domestic consumption of hot water is 45 L/person a day at 60°C. This works out to 3400 MJ/person a year or about 1000 kWh/person a year given that the water is heated initially from 10°C. About 50% of this energy can be provided by the solar collectors. So for gas heated boilers rated at 83% efficiency 500 kWh/person a year is translated to 600 kWh which produces 115 kgCO$_2$/person a year; and using solar thermal collectors, therefore, saves 115 kgCO$_2$/person a year.

For electrically heated domestic water (the electricity being currently sourced from a mix of fossil fuel and renewable energy) efficiency at the consumers' meter is 100% and 500 kWh/person a year produces 210 kgCO$_2$/person a year with 210 kg of CO$_2$ saved by the solar collectors. Using electricity for domestic hot water, therefore, shows an increase in carbon dioxide emissions of 84% over the use of natural gas.

Summary

The reader may like to verify the figures given above; they are calculated from readily available data. Solar thermal collectors are not new and have been in use for many years. However, they have not been popular largely due to cost and the long payback. This situation will change following legislation and with future dramatic increases in the price of fossil fuel.

A scenario for a large community

The principle of total reliance upon the utilities of water supply, drainage, electricity supply and gas supply to every building or community is now being challenged. Distribution networks suffer from the need for ongoing maintenance, updating, replacement and losses, as well as disruption.

The move to local heat and electricity generation may not provide a totally self-sufficient site and the need for the National Grid and natural gas utilities is still there. However, there are examples of sites that are even more self-contained. One well known example is the BedZED development in Beddington, Surrey.

The following offers one probable solution for a largely self sufficient and sustainable site.

Water. There are usually three sources of water available to a site: rainwater, groundwater and treated mains water. Rainwater can be collected from the roofs of the buildings on site, stored and used for flushing toilets. A borehole that taps into an aquifer can supplement the rainwater for non-potable uses. Low water use appliances can be used to reduce water demand. These systems can reduce reliance on mains water by 40%.

Organic waste. Non-organic waste will rely on an efficient recycling facility provided by the local authority or their sub-contactors. Organic waste, such as that from kitchens, can be fed into an anaerobic biodigester where it is used to create biogas or methane. Sewage from effluent will also feed the digester and the resulting gas will be mixed with natural gas to power the combined heat and power plant. Treated effluent from the digester will disperse into the ground and the inert digestate will provide soil compost for use on the site.

Combined heat and power. A mixture of biogas and mains gas fuels the combined heat and power (CHP) plant. The products from which are augmented by the use of flat plate or evacuated tube solar thermal collectors serving a thermal store for top up space heating in the winter and top up hot water supply throughout the year. Photovoltaic arrays and wind turbines will contribute to the power supply from the CHP unit.

Vapour absorption cooling. If air conditioning is a requirement for the site, an absorption chiller can be used. This requires heat to operate and therefore can utilise some of the heat generated during the summer by the CHP plant. It can be supplied to the cooling plant in any convenient form such as electricity, steam, solar energy or waste heat. Two common types of absorption cooling equipment available for use are ammonia/water in which the ammonia is the refrigerant and the water is the absorbent, and lithium bromide/water in which the lithium bromide is the absorbent and the water is the refrigerant. The latter is more suitable for chilled water air conditioning requirements in buildings.

The addition of absorption cooling to the CHP plant is known as tri-generation and further increases its fuel efficiency.

A4.8 Further reading

Capitalism as If the World Matters, J. Porritt, *Earthscan*, 2005.
Building Envelope, P. Wakefield, *EIBI*, January, 2006, CPD Collection.
Ground Source Heat Pumps, G. Maidment, *BSJ*, January, 2006, CPD Collection.
Sustainability, A. Pearson, *CIBSE BSJ*, January, 2006.
Solar Thermal, T. Dwyer and B. Sinclair, *BSJ*, February, 2006, CPD Collection.
Air Source Heat Pumps, T. Dwyer, *BSJ*, April, 2006, CPD Collection.
Lighting in Commercial Buildings, G. Mountford, *EIBI*, April, 2006, CPD Collection.

Sustainable Resources, A. Rowley (director), Resource Efficiency Knowledge Transfer Network. Defra. Environmental and Sustainable Management, May/June, 2006.
The Future in Numbers: 50 Technologies, *CIBSE BSJ*, Special supplement, August, 2006.
Using Thermal Mass to Reduce Energy Consumption, T. Dwyer, *BSJ*, October, 2006, CPD Collection.
Underfloor Heating and Renewables, J. Goth, *EIBI*, January, 2007.

A4.9 Appendix closure

Sections A4.1, A4.2 and A4.3 make for important reading in order to understand the urgent need for sustainable building. These sections are particularly appropriate for building engineers who hopefully have already been introduced to the thermodynamic laws (of nature) and their relation to the built environment.

This appendix is not intended as an exhaustive treatise on sustainable building engineering. The concepts that provide the encouragement for the sustainable approach to building and building services have been introduced. It has been shown that the whole life cost for a building is between three and five times the initial capital cost and some building owners are taking the opportunity to account for this by ensuring that building products and systems are selected for their whole life cost rather than the lowest purchase price.

It is a transforming experience because it changes the whole attitude in procurement practice philosophy and one in which architects, builders and building services engineers have a major role to play.

Bibliography

A.F. Burstall, *A History of Mechanical Engineering*, Faber, 1963.

D.V. Chadderton, *Air Conditioning, a Practical Introduction*, 2nd Edition, E&FN Spon, 1997.

J.F. Douglas, *Solutions to Problems in Fluid Mechanics*, Part 1, Pitman, 1975; Part 2, 1977.

O. Fanger, *Thermal Comfort*, Macgraw Hill, 1972.

J.A. Fox, *An Introduction to Engineering Fluid Mechanics*, Macmillan, 1977.

S. McLean, Natural Ventilation of Buildings, *EIBI*, March, 2006, CPD Collection.

Memmler, Cohen and Wood, *The Human Body in Health and Disease*, Lippincott, 1992.

K.J. Moss, *Heating and Water Services Design in Buildings*, 2nd Edition, E&FN Spon, 2003.

K.J. Moss, *Energy Management in Buildings*, 2nd Edition, Taylor & Francis, 2006.

J. Porritt, Capitalism as If the World Matters, *Earthscan*, 2005.

G.F.C. Rogers and Y.R. Mayhew, *Thermodynamic and Transport Properties of Fluids*, Blackwell, 1995.

J.R. Simonson, *Engineering Heat Transfer*, Macmillan, 1981.

The Chartered Institution of Building Services Engineers, *The CIBSE Guide Books A, C and G*.

The Institute of Plumbers, *The Plumbing Guide*.

Index

Note: This index should be read in conjunction with the chapter contents and the full list of examples and case studies in the front of the book.

Absolute roughness values 161
Absorptivity 73, 75
Adenosine triphosphate 294
Air flow through openings 191
Air pollution zone 190
Air pressure zones 186–9
Air temperature 11
Air velocity distribution 113
Ambient temperature 18
Anaerobic biodigester 302

Bellmouth 147
Bernoulli equation 100, 137, 237, 273
Biodiesel properties 284
Biofuel 284
Biomass calorific values 283
Biomass CO_2 emissions 283
Black, grey and selective surfaces 78
Black body emissive power 78
Black surface 76
Body heat loss 6, 7, 9
Boiling and condensing 212
Boundary layer formation 122
Boundary layer separation 124
Box's formula 120, 180, 236–7
Boyle's law 182, 183
Buckingham's Pi theorem 242
Building stock, UK 297

Capacity ratio 220, 221
Carbon footprint 300
Cell biology 293
Chezy formula 170
CHP additional features 287
CHP assessment in buildings 286

CHP heat to power ratios 287
CHP plant 286
Clo 17
Coefficient of discharge 106
Colebrook & White formula 126
Collection (solar thermal) ability 279
Collector (solar thermal) types 279
Colour temperature guide 76
Comfort scales 22
Comfort temperature 13
Community services 301
Compressed air pipe sizes 183
Compressed air pressure factor 181
Continuity of flow 99
Crimp and Bruge's formula 177
Crossflow LMTD correction 226
Cross ventilation through openings 196

Dalton's Law 12
D'Arcy's formula 116, 236, 240, 242, 244
D'Arcy-Weisback formula 177
Dew point temperature 12
Dimensions of terms 235
Dynamical similarity 243

Ecological footprint 300
Economic growth 291
Embodied energy in building materials 298–9
Emissivity 73, 75, 76, 81, 279
Effectiveness 220
Energy dispersal 293
Entropy increase 296

Index

Environmentally friendly building materials 298
Environmental temperature 14

Forced convection 55, 56, 58
Form factors 80
Fouling factors 213
Fourier's law 48
Francis turbine 268
Free convection 55, 58, 62, 63
Frictional losses 146
Fuel cell, types 288
Fuel cell efficiency 289

Geometric similarity 243
Globe thermometer 12, 13
Grashof number 61, 63, 90, 256
Grey surfaces 76

Hagen's formula 116
Heat exchangers in use 209
Heat flow paths 30, 31, 33, 36
Heat flux 29
Heat radiation wave form 77
Heat transfer coefficient 70, 71
Hydraulic gradient 154
Hydraulic mean diameter/depth 170
Hydroelectric system 270
Hyperpyrexia 5, 7
Hypothermia 5

Impermeability 173
Incident heat radiation 75

Kaplan turbine 267, 269
Kirchoff's law 78

Lambert's law 79, 83
Laminar flow characteristics 118
Laws for pumps and fans 246
Log mean temperature difference (LMTD) 59, 60, 204, 211

Manning formula 176
Manometers, inclined differential 103
Material waste 293
Mean radiant temperature 12
Mean spring peak water velocity 276
Mean velocity 100
Medulla oblongata 10
Metabolic rate (Met) 7, 17

Methane release 283
Moody chart 127

Number of transfer units 220
Nusselt number 253, 256

Organic waste 302

Pelton impulse turbine 267, 268
Phase change materials 298
Photosynthesis 292, 293
Photovoltaic systems 282
Plane radiant temperature 96
Planet's wellbeing 292
Plank's law 79
Poiseulle's formula 116
Pole's formula 180, 261
Power, dimensions of 264, 267
Prandtl number 253, 256
Predicted mean vote 21
Predicted percentage dissatisfied 21
Pressure factor for compressed air 181
Pumped storage system 272

Rainwater run-off 173
Reflectivity 74, 75
Reynold's experiment 117
Reynold's number 242, 244, 251, 253

SEDBUK rating 299
Selective surfaces 76
Sewage use 302
Shock losses 146
Soil stack flow 172
Solar constant 278
Solar intensity 278
Solar thermal collectors, application of 300
Spectral proportions of heat radiation 77
Static loss of rising air in a vertical duct 157
Steady flow 99
Stephan–Boltzman constant and law 79
Stephan's law 78
Submersible pump 139
Suction lift 139
Surface characteristics 75
Surface conductance 71
Surface roughness 126

Temperature scales 3
Thermodynamics, the zeroth law 294
Thermodynamics: the first law 295;
 the second law 295; the third
 law 296
Turbulent flow characteristics 120

Uniform flow 99

Vacuum lines, classification in Torr 183
Vapour absorption cooling 302

Vasoconstriction 10, 24
Vasodilation 10, 24
Vector radiant temperature 96
Velocity pressure loss factors 151
Venturi, inclined 110

Water resources 291
Wein's displacement law 77, 79
Wet bulb temperature 11, 14
Window J factors 199

Printed in Great Britain
by Amazon